Acclaim for

Harvest for Hope

"Proposes actual solutions . . . Give HARVEST to someone you love."

—*Seattle Weekly*

"I love this book! Jane Goodall's generous, playful spirit imbues every fascinating page. HARVEST FOR HOPE is full of mind-expanding observations . . . a personal, tender wake-up call telling us that we can reclaim the wisdom of our bodies."

—Frances Moore Lappé, author of *Hope's Edge* and *Diet for a Small Planet*

"An excellent resource . . . Jane Goodall, as always, offers hope."

—*Satya*

"Jane Goodall convinces us that we should have a new relationship with food, one that is inspiring and delicious, at the same time a preservation of tradition and an act of conservation."

—Alice Waters, author of *Chez Panisse Fruit* and *Chez Panisse Vegetables*

"Read this book for clear, logical arguments from an individual known for her wisdom and humanity . . . She provides many specific suggestions of what individuals can do . . . There's plenty of both disturbing and inspiring food for thought."

—BookLoons.com

"If you haven't thought much about the food you eat and the choices you make (and even if you have), this is an important book to read!"

—Deborah Madison, author of *Vegetarian Cooking for Everyone* and *Local Flavors: Cooking and Eating from America's Farmers' Markets*

more . . .

"A lucid, anecdote-filled introduction to the world of food, revealing how our food production affects us and how our choices affect the environment . . . Consider this book the shopping list for you and your children's future."

—Paul Hawken, author of *Ecology of Commerce*

"If you want to be newly awakened to the joy of eating, to the miracle of food, and to the power each of us has by the way we live our lives, do yourself a favor. Get a copy of HARVEST FOR HOPE. I promise you: Your life will change in countless ways, all of them for the better."

—John Robbins, author of *The Food Revolution* and *Diet for a New America*

"Thrice a day you get the chance to change the planet. You can change it in significant ways, if you follow just some of this book's wise advice."

—Bill McKibben, author of *Wandering Home: A Long Walk Across America's Most Hopeful Landscape*

"An inspiring book on the impact humans' food choices have on the planet . . . Goodall has always been prophetic. She has consistently lived and written from her heart . . . Slow down, Goodall advises. Pay attention. Be mindful. Take the right kind of bite and help save the world." *—Ms. Magazine*

ALSO BY JANE GOODALL

Reason for Hope

Harvest *for* Hope

A Guide to Mindful Eating

Jane Goodall

with Gary McAvoy
and Gail Hudson

WELLNESS
CENTRAL

NEW YORK BOSTON

The table on page 156 is by Rich Pirog, 2002. "How Far Do Your Fruits and Vegetables Travel?" Leopold Center for Sustainable Agriculture, Iowa State University. From the In-a-Nutshell series "How Far Does Your Food Travel?" from the Center for Urban Education about Sustainable Agriculture.

The Farm Sanctuary Adopt-a-Turkey Project information on page 113 courtesy of Farm Sanctuary.com

Wellness Central
Hachette Book Group
237 Park Avenue
New York, NY 10017

Visit our website at www.HachetteBookGroup.com.

Printed in the United States of America
Originally published in hardcover by Hachette Book Group.
First Trade Edition: September 2006

10 9 8 7 6 5 4 3

Wellness Central is an imprint of Grand Central Publishing.
The Wellness Central name and logo are trademarks of Hachette Book Group, Inc.

The Library of Congress has cataloged the hardcover edition as follows:
Goodall, Jane.
Harvest for hope : a guide to mindful eating / Jane Goodall with Gary McAvoy and Gail Hudson—1st ed.
p. cm.
Includes bibliographical references.
Summary: "An exploration of the global meaning of food and what all of us can do to exercise power over the food industry and, ultimately, our environment"—Provided by the publisher.
ISBN 0-446-53362-9
1. Gastronomy. 2. Food industry and trade. 3. Environmental protection. 4. Food. 5. Nutrition 6. Diet. 7. Organic Farming. I. McAvoy, Gary. II. Title.
TX631.G66 2005
641.3'002—dc22

2005018832

ISBN 978-0-446-69821-4 (pbk.)

Cover design by Anne Twomey
Book design by Giorgetta Bell McRee

This book is dedicated to the thousands of small farmers who are valiantly struggling to survive, especially those who have embraced organic practices; to those who stand up to speak out against the bullying tactics of agribusiness; to the men and women who work tirelessly to reintroduce the citizens of the fast food nations to real, wholesome food.

And to the billions of farm animals held in torment around the world.

Acknowledgments

This book would never have been started but for Gary McAvoy, who convinced me that such a book was needed, and that my voice should deliver the message. It has involved endless research, talking with people, and, most important, thinking through the issues. And none of this would have been possible without the help and generosity of many people along the way.

I worked to finish the text of this book during my May to June 2005 U.S. tour, and I could not have coped without the help of Mary Lewis and Rob Sassor. Mary must have sent off thousands of e-mails to distant places around the globe, often working late into the night. Rob was traveling with Mary and me, and he drafted much material, followed up leads, and worked tirelessly to help us meet the deadline.

In writing the chapter dealing with genetically modified food, in particular, I sought advice from many people: Tom Griffith-Jones; Lord Peter Melchett; and Dr. Stanley Ewan, who, working with Dr. Arpad Pusztai, provided valuable insights regarding the work with rats and genetically modified potatoes and tomatoes.

Howard Buffett helped me to understand a variety of issues regarding the growing of corn, and Robert Eden of Comte Cathare vineyards in France told me a lot about his organic vineyards. I am thankful to Nancy and Jerome Kohlberg of Cabbage Hill Farm, who shared their insights on organic farming. Russ Miller and Phillip Evans of Turner Enterprises offered their expertise on sustainable ranching. And I am grateful to Shadowhawk and his son Washo for sharing their beautiful story with us.

Katie Faulkner helped us to get a number of photographs from Compassion in World Farming, and Margaret Foster sent some from Advocates for Animals. Lesley Day, of Chimps, Inc., provided photos and valuable research. Gloria Grow of Fauna Foundation also helped with photos. In particular, my good friend Tom Mangelsen not only provided many photos but spent considerable time obtaining images for this book. My deepest gratitude to you, Tom.

Many people worked incredibly hard, sometimes into the night, to select our photographs, get them in the right format, and get permissions. My heartfelt thanks to this team: Jeff Orlowski, Rob Sassor, Mary Lewis, Brent Wenner, and most particularly, Michael Aisner and Nona Gandelman, who worked all through Independence Day. I also want to thank Emily Griffin at Warner Books, who received our images from all over the world.

Judy Waters, my sister, kept my body and soul together with her great (organic) cooking during the time I was able to write at our home, the Birches, in the U.K.

I am grateful for the attention and many helpful suggestions given to the manuscript by Natalie Kaire of Time Warner. Without Gail Hudson, who joined us half way through this project, *Harvest for Hope* would not have been ready for publication. Not only does Gail work extraordinarily hard, but she has a passion for the subject and she writes like a dream. She has been a delight to work with, I am thrilled she came on board—and I have made a new friend.

I am eternally grateful to my wonderful friends Jonathan Lazear, our agent, and Jamie Raab, our editor at Warner Books. I cannot imagine many agents or publishers who would have shown such patience and understanding. We thank both of you so, so much.

Jane Goodall, July 2005

Many thanks to each of the friends and colleagues who shared their ideas, inspiration, and help in shaping this book: Renee Bell, Marlene Blessing, John Burgess, Richard Chadek, Hans Cole, Joy Delf, Mike and Virginia Duppenthaler, Charles and Rose Ann Finkel, Deborah Koons Garcia, Kathy Humphrey, Robin Kobaly, Tara Kolden, Laura Krebsbach, Frankie Lappé, Deborah Madison, Dave Miller, John Mullen, Mary Ann Naples, Boanna Owens, Brenda Peterson, Clarice Swanson, Doug Thompson, Peter Vartabedian, Craig Winters, and Sue Conley and Peggy Smith at Cowgirl Creamery.

Gary McAvoy and Gail Hudson, July 2005

Contents

Introduction

Why, people ask me, are you writing a book about food? I suppose it seems strange to those who do not know me: I am, after all, the "Chimpanzee Lady"; in Asia I am often called the "Chimpanzee Mother." So why would I be interested in eating? Let me explain.

I spent hours and hours, from 1960 onward, watching chimpanzees feeding. I collected specimens of what they were eating. I noted the behavior associated with eating—keep everyone away, if you possibly can, unless there is more than enough for everyone. Which means, if you are the boss, you can keep everyone away from food that you really enjoy. I found, over time, that dominant females had a more successful reproductive career—they started having babies earlier, and they had more of them. So it was clear that high dominance rank helped you get more food, and the best food, and that this would help your offspring, especially your daughters. Then they, like their mothers, would become high-ranking and successful. Food, therefore, was important. And, I found, they would fight over food, especially a favored food in short supply.

When I was not in the forest, where I most loved to be, I spent a lot of time people watching. Just for fun. In the late 1960s, when I began giving lectures about chimpanzees, I was really terrified of public speaking (though no one knew it). I often found myself at dinner parties that took place before I spoke. With my stomach in knots, there was no way I could eat. So I used to watch people. It helped calm my nervousness because, underneath the supposed civilized veneer of twentieth-century *Homo sapiens* I could so easily see the same kind of behavior I had, for so long, watched in my chimpanzees.

Then, in 1986, I attended a Chicago Academy of Sciences conference called "Understanding Chimpanzees" that changed my life. All those studying chimpanzee behavior in Africa, and many studying them in zoos, came together, under one roof, for a four-day conference. It became clear that these amazing apes were in trouble. Their habitat was vanishing. They were treated cruelly in many captive situations. And they were being eaten. They were being hunted and sold for food to make money. I found that this had become a serious problem—not only for chimpanzees, but for all forest animals—because, in addition to subsistence hunters who have from time immemorial lived off the bounty of the forest, animals were now being hunted commercially. This was made possible by the new roads built by foreign timber companies, which provide access to previously inaccessible forests. Hunters rode the trucks, camped at the end of the roads, shot everything edible, from elephants to bats, smoked it, and sold it in town, where the urban elite paid more for this "bushmeat" than for chicken or goat meat.

I left the conference knowing that I had to do what I could to help save chimpanzees, improve their condition, pay back something of what they had given me. I knew I could no longer sit and observe them in the forests I love.

Instead I had to travel the world, raising awareness of their plight. For they are on the brink of extinction.

It was not long before I realized that the problems facing the chimpanzees are closely connected with the problems facing Africa. And I soon came to understand that, in many cases, these problems could be laid directly at the door of the unsustainable lifestyle of the elite societies around the world. A lifestyle hatched in the Western world and exported, along with its values (or lack of them) and its technology, to the developing world. To help my chimpanzees, therefore, I had to start thinking about ways of opening the eyes of those who were, unknowingly, robbing the natural world of more and more of her nonrenewable natural resources.

There are so many, many ways in which we are destroying the planet. And once we understand, once we care, then we have to do something. When I came to this realization, I knew I had a unique perspective. I had grown up in a time of austerity—during World War II—and learned to take nothing for granted. I realized that having enough—of clothes and of food—was what mattered. And I went from that happy childhood to the forests of Africa to study the chimpanzees and live my dream. When I first arrived in Gombe National Park of Tanzania, I found a world where all was pure—the springs that gave rise to the streams of Gombe were nurtured deep in the heart of an uncontaminated watershed. There were no man-made chemicals in the forest. Lake Tanganyika was the largest body of unpolluted fresh water in the world.

And then, gradually, it all changed. The population living in the forests around Gombe multiplied. There were refugees from Burundi and the Congo. The trees were cut down. Soil erosion laid bare the once verdant mountain slopes. The people—peasant farmers and fishermen, the poorest of the poor—increasingly suffered. In their struggle to survive they cut down the trees and the soil washed away. Many were hungry. Well-meaning foreigners interfered with the ancient fishing methods that had enabled the fishermen to live in harmony with their world, so that in the end Lake Tanganyika was overfished. The people became increasingly poor, increasingly hungry.

And I would leave Tanzania to lecture in Europe and America and see people eating, always eating. More and more food bought, more and more thrown away. People dying of eating too much while the people of Africa, whom I had just left, were starving. I couldn't try to help only the chimpanzees when people were struggling to survive. So it became clear: To help the chimpanzees it was necessary to work with the people living in the villages around Gombe.

Gradually I moved into different circles, where I began to learn more about poverty. And hunger. As I traveled further and further around the world, I met more and more young people who had lost hope. I found despair, apathy, and anger. And I came to understand, listening to the voices of the wise, that humans are following a path that could easily lead to the end of life on earth as we know it. For we are polluting, with synthetic chemicals, the air, the water, and the land. And a huge amount of that pollution is caused by the agricultural chemicals that are used as fertilizer, pesticide, and herbicide to grow our food. Chemicals that, in some instances, were developed for use in the war, against human enemies. I found, too, that we are guilty of unbearable cruelty to the animals we raise as food. We have so far to go to realize our human potential for compassion, altruism, and love.

It has become increasingly clear that growing, harvesting, selling, buying, preparing, and eating food plays a central role in the world. And it is equally clear that some things are going wrong. Much of our food is unhealthy. Many people are no longer aware of where their food comes from. Some have *no* idea what they are eating. In fact, over the past hundred years—especially during the half century since the end of World War II—the industrial, technological world has increasingly destroyed our understanding of the food we eat: where it comes from and how it reaches our tables.

At one time humans had a more intimate connection with the earth and the animals that sustained them. They collected their food, used primitive tools to crack open nuts, dismembered a carcass after a successful hunt, and, when they had found fire, gathered to eat in their rock shelter—probably with hyenas or wolves waiting for scraps outside. And then, with the coming of agriculture, people worked the fields, plowing and sowing, and then harvesting. Everyone joined in, helping to bring in the grain and hay before it rained. And the women baked and boiled and fried the food, and hung hams from the ceiling, and cows were milked, and butter and cheese was made in the dairy. The seasons were important. And the weather mattered because of how it affected the crops.

Today there are more than six billion human beings on this planet, all needing to eat, and we have seen the rise to power of giant multinational corporations, eager to exploit the opportunity of such a vast global market. In order to produce more and more food, either to feed hungry people or to satisfy the *wants*—as distinct from the *needs*—of today's materialistic urban elite, agricultural methods have changed. The aim of the ever more powerful corporations, with the support of governments they help to bring to power, is to produce as much food as possible, as cheaply as possible, with maximum profit for the shareholders.

We live in an era when the poisoning of earth, water, and air by agricultural chemicals is making people, animals, and the environment sick—even creating diseases. An era when tropical rain forests are clear-cut to grow corn to feed to cattle. When animals raised for food are farmed more and more intensively without space or dignity and fed rich, often unnatural, foods to ensure that they put on as much weight, or produce as much milk, or lay as many eggs as possible, in the shortest time, for the greatest profit per animal per day.

Most people have no idea that these corporations control increasing areas of farmland around the world along with increasing numbers of the seeds from which our food is grown. And that they control how the seeds are grown—in monocultures in huge, poisoned fields. Most people don't know that the corporations have taken over the production of meat. Or that they are gradually forcing out the last of the traditional small family farmers. Nor have most people realized the speed with which multinational corporations are taking over local grocery stores that once sold local produce. Indeed, many of our regional foods, our rich diversity of crops, are now considered endangered because of the corporate control over our food and culture.

This has created a barrier between us and the land and our food, a barrier that is supposed to prevent us from realizing the devastation and suffering that, so often, is implicit in each bite. Today more and more people in urban centers around the world can buy a frozen precooked dinner from a grocery store, or a meal in a restaurant, and eat the food with no knowledge of what it really is, how it was grown or harvested or cooked, or where it came from. Most people never even wonder how far the food they eat has traveled to get to their local grocery stores, and how much energy and resources was used to get it there.

What can be done to stop the monstrous corporate takeover that is motivated by the desire for more, and ever more, financial profit? How can we change a world in which decisions that affect our health and the health of the planet for generations to come are taken in order to present economic growth at the next shareholders meeting? What can we do as individuals in this world of giant corporate greed, human and animal suffering, and environmental destruction? Is it absolutely beyond our control? There are millions of people who think this way, who feel helpless in the face of the magnitude of the problems. So even if they care they fall into apathy. This book is a clarion call to rouse such people from spineless acceptance of the status quo. I cannot stress strongly enough that *every individual makes a difference.* And I hope that you will understand just what it is that you can do—and that you choose to do it!

I shall introduce you to some wonderful people who are fighting back, tackling the corporate giants—and, as with David and Goliath, sometimes coming out on top. They are truly inspirational, illustrating the indomitable human spirit. And in less dramatic ways, thousands more are doing their bit. More and more frustrated and alarmed consumers all over the world are refusing to buy from fast food chains and are insisting on organic. Every day if all of us who can afford to, make ethical choices as to what food we buy and eat, and from whom we buy it, we can, collectively, change the way our food is grown and prepared.

I hope, in this book, that I can increase understanding of vitally important issues—important to the sustainability of the natural resources of the planet, the well-being of animals, and, very importantly, human health.

It is not too late to change direction. We can once again become connected with the food we eat, and learn to understand its nature and its history and embrace a more natural diet. We must, for we are at a critical point in human history. If we continue to allow the corporate world to control our food supply, we could devour or poison, within the next half century, all the food resources that sustain us.

I have three little grandchildren and when I think of how we have harmed this planet since I was their age I feel deep pain. For their sake it is so important to reverse the trend toward destruction. And one way in which we can truly make a difference is to think about what we eat. Each decision we make—what we choose to buy, choose to eat, will have an impact on the environment, on animal welfare—

and, importantly, on human health. And so, understanding this, I decided it was important to write a book about it. A book that will, I hope, help people understand what is going on, so that everyone understands the important role each one of us can play in making this a better world. It will be tough going in parts—we truly have made a mess of things. But I hope, too, it will be fun. and that it will bring hope for the future.

Together let us join hands and hearts as we each do our bit to create a better world for our children, and theirs. So that the harvest we leave for them to reap will truly be a "harvest for hope."

Chapter 1 Animals to Us

"In the cosmos there are only eaters and the eaten. Ultimately, all is food."
—Hindu Upanishads

There is an old English saying, "Manners Makyth Man." In fact, it is food that makyth man. For if we discount our basic physiology and anatomy, and those behaviors inherited through our genes, we truly are what we eat. We cannot survive without food, although certain individuals have claimed that we can. In the second half of the nineteenth century, young fasting girls captured the imagination of people all over Britain, especially one Sarah Jacob, a twelve-year-old who confounded doctors by refusing to eat or drink for two years. She became a tourist attraction, and in the end her parents agreed that a medical team could keep her under surveillance. It did not take long for her condition to deteriorate and soon she was dead. Her father, who had refused to offer her food, was sentenced along with his wife for manslaughter. It was never made clear who had been secretly providing her with sustenance during the previous two years.

More recently, in 1999, Ellen Greve set herself up as a

New Age dietary guru who claimed that she had not eaten for five years, surviving instead on invisible crystals in the air. She advocated a twenty-one-day fasting regime linked with spiritual exercises, the details of which she sold to her more than 5,000 followers—it made some very sick and killed three of them. Eventually she was challenged and, with her reputation at stake, she agreed to stay in a hotel room where all comings and goings could be monitored. After three or four days, during which she definitely ate nothing, she collapsed and was admitted to the hospital. Unlike poor Sarah, she recovered, and tried to save face by complaining that the air in her hotel room was not fresh and could not sustain her as did the air in the Australian Outback. To which she fled—and I have not heard of her since!

It is true, of course, that people can fast for a surprisingly long time, but in the end everyone needs some kind of nourishment—it can be a surprisingly small amount—to stay alive. All creatures need food of some sort, although some can fast for much longer than humans can. Hibernating species like bears enter a state of slowed-down physiological processes and can survive a long harsh winter without feeding at all. The African lungfish buries itself in the mud of a drying waterhole, and waits it out until the next rains—which can be several years. A tick that I met was still alive after living in a jar, without food or water, for more than six years. It got very excited if you put your hand near and frantically waved its front feet and antenna—I suppose it sensed blood and I felt sorry for it. But these are exceptions. Most animals, like us, need food—and especially water—on a much more regular basis. And Planet Earth provides a staggering number of items on the menu—almost everything is food for someone or something. In nature there are thousands of fascinating stories, plots and counterplots, that revolve around this basic need for food.

NATURE'S INGENUITY

Some animal species have evolved the most extraordinary ways to find, catch, prepare, or digest the substances, plants, or other animals that keep them alive. Live prey is chased, stalked, poisoned, snared. Spiders are incredibly skilled in enmeshing, trapping, or hunting their prey. There is even one spider who puts a sticky blob on the end of a short strand of web and then, like a little cowboy with a lasso, waves it around her head at passing flies. The archer fish waits for a fly to land on a branch overhanging the stream, then launches a mouthful of water with deadly aim, knocking his dinner into the water. Ant lion larvae dig funnel-shaped pits in loose sand, lie in wait at the bottom, and hurl grains of sand when they sense some hapless insect struggling at the rim, causing it to lose its footing and skid down to be seized in strong mandibles. Many creatures, once they grab their prey, inject poison that incapacitates their victim. This enables them to eat creatures larger and stronger than themselves. There is even a family of plants that lives on animal foods. Pitcher plants lure insects into pitcher-shaped leaves filled with a delectable enzyme soup in which the prey is gradually digested and absorbed. The sundew has sticky leaves that close shut over unwary insects stopping there to feed on tempting drops of nectar—they are then gradually digested.

Different kinds of animals use different structures and methods to achieve a similar goal. To reach the nectar buried deep within a flower, insects such as butterflies and bees use a long proboscis; hummingbirds and sunbirds use long thin beaks. To feast on succulent termites or ants that are hidden underground, armadillos and anteaters have evolved strong digging claws and long, wormlike sticky tongues that snake down into the mound; chimpanzees fish for them with straw tools. Elephants reach food high up in a tree with their trunks, giraffes make use of their long necks, other creatures get up there by climbing or flying. Animals as diverse as spiders and lions hunt by stealth, stalking their prey, or hiding and pouncing. Others, like cheetahs and falcons, rely on short bursts of speed or, like hyenas, show amazing powers of endurance in the chase. Prey may be located by sight, hearing, scent, vibration, or echo location.

In turn, many plants and animals have taken equally ingenious measures to protect themselves from becoming food. Insects, according to their species, have evolved to resemble the bark of trees, dead leaves, flowers, twigs, and so on. There is a caterpillar at Gombe that looks exactly like a bird dropping. Caddis fly larvae make themselves little tubes to live in, then camouflage them by sticking on bits and pieces of the surrounding vegetation. Trapdoor spiders use the same technique to disguise the hinged plugs that close the entrance to their burrows. Some insects have brilliant colors but taste revolting. After one experience the would-be diner will avoid others of their kind forever. And there are other insects that actually taste delicious but that have evolved to resemble the noxious bugs so closely that they are avoided!

Many of the larger herbivores are arrayed in spots and stripes that blur their outline and make them hard to see. Octopuses can even change color, the better to blend into their surroundings. A diverse collection of creatures—porcupines, hedgehogs, puffer fish, sea urchins, hairy caterpillars, and so on—protect themselves with spines, quills, prickles, or stinging hairs. Others develop a tough outer armor, like tortoises and turtles, armadillos, and countless insects. There are some creatures who have developed poison, administered through the teeth, as in snakes, or stings as in the cone shells and stonefish, that, while primarily used to immobilize prey, is also extremely effective in repelling would-be predators. The same is true for the electric shocks administered by stingrays, electric eels, and so on.

As for the plants, they and their seeds are protected in a

hundred different ways, by thorns and prickles and stinging itching hairs and foul toxins, and hard outer coverings. Many plant products, however, are meant to be eaten. Succulent fruits have been designed as high-quality food so that fruit-eating animals are more than happy to play a role in seed dispersal, carrying them away in their stomachs for eventual excretion elsewhere. Some seeds cannot germinate until they have passed through the stomach and guts of an animal. Many plants have developed alluring fragrances to attract insects, certain birds—and even a species of bat—to feast on the sweet nectar secreted in their flowers. These gourmets transport pollen from one plant or tree to another, thus playing a vital role in the propagation of the species.

Internal organs and digestive systems have adapted to cope with all manner of foods: tough, fibrous vegetable matter, leaves full of toxins or covered with spines, putrid carcasses, bones, and so on. Jaws and teeth of different size and strength enable their owners to crush, tear, or chew whatever it is that nature has planned as their diet. Birds are equipped with a fascinating array of beaks, each designed for dealing with the food the bird is born to eat. Hyenas have teeth and jaws so strong that they can crunch big bones and digestion so incredible that they can extract some nutrients from ancient carcasses.

By and large animals can only eat what they were born to eat: A giraffe could not survive on meat any more than an eagle could survive on leaves. Many species are quite specialized in their dietary needs: Koala bears must have eucalyptus leaves, giant pandas need their bamboo, the larvae of hunting wasps can only survive when fed the paralyzed bodies of particular species of spider or caterpillar. Other creatures are more catholic in their tastes, and many are omnivores, surviving on a mixed diet of plant and animal foods.

Thus, to a large extent the structure and behavior of animals has been determined, during the course of evolution, by their need to get adequate food of the right sort. And there can be little doubt that food—its acquisition, preparation, and consumption—played a role in the evolution of our own species. Like many of our primate relatives, we humans are omnivores. So are chimpanzees, from whom we differ genetically by only about one percent. Many people are interested in chimpanzee diet for the insights it may give us into the food preferences of our stone age ancestors. Chimpanzees are primarily fruit-eaters—they have long mobile lips and special ridges on the insides of their cheeks that enable them to suck and squeeze the juice from their food. But they also eat leaves, flowers, and stems, as well as leaf buds, seeds, and nuts that are rich in vegetable protein. They enjoy animal protein, too, and at certain times of year consume large numbers of insects, primarily ants, termites, and caterpillars. And at intervals throughout the year they hunt small- and medium-sized mammals; meat makes up about 2 percent of their annual diet at Gombe.

TOOL USE AND HUNTING

For anthropologists with an interest in human evolution, such as Louis Leakey, the most significant observations that I made at Gombe, in the early 1960s, were those that documented, for the first time, chimpanzee tool-using and hunting behavior. I can never forget the first time I saw a chimpanzee using tools. I was trudging through wet vegetation after a frustrating morning—for most of the chimpanzees were still shy, running off whenever they saw me. Suddenly I saw a black shape squatting by a termite mound. Peering through the leaves I saw that it was David Greybeard, the male who was beginning to lose his fear of the strange white ape that was me. I saw him pick a grass stem, poke it down into the mound, wait a moment, then pull it out coated with termites. These he picked off with his lips. I could see his jaws working and hear the sound of scrunching. I had seen a wild chimpanzee using a tool!

It was such an exciting observation that afterward I almost thought I must have imagined it. But a few days later I saw both David Greybeard and his friend Goliath using grass stems to feast on termites. And I watched as David broke a leafy stem from a nearby bush and stripped off the leaves—modifying an object to suit it to his purpose. Not only had I observed a wild chimpanzee using a tool but I saw one actually making a tool! Back then it was thought by scientists that only humans used and made tools. This, it was held, differentiated us from the rest of the animal kingdom more than any other criteria. "Man the Toolmaker" is how we were described in the anthropological textbooks of the time. I sent a telegram to Louis Leakey. "Well," he replied. "Now we must redefine Man, redefine Tool—or accept chimpanzees as humans!" Subsequently I would observe the chimpanzees using long peeled sticks to feast on army or driver ants, scrumpled-up leaves to sop water from hollows in tree trunks, and a variety of other objects for different purposes—most of which were in the context of obtaining food.

It was David Greybeard who provided me with the first evidence that chimpanzees sometimes eat meat—prior to my study it had been assumed that chimpanzees were vegetarians. On that first occasion I saw David feeding on a baby bush pig. He was sharing the flesh with an old female who sat close, begging, while her child, unsuccessful in getting a share from her elders, made repeated sorties to the ground to snatch up scraps. She was charged each time by the infuriated adult pigs and had to rush back up the tree, screaming loudly. A few weeks later I actually watched a successful hunt. A small group of red colobus monkeys had taken refuge at the very top of a tall tree that rose up out of the canopy. This was a mistake, for in such a situation they are relatively easy to catch. Several adult male chimpanzees took up positions in the surrounding branches, effectively cutting off the monkeys' escape routes. Then an adolescent male climbed slowly up the trunk, leaped toward a female monkey who had an infant clinging to her breast, seized the baby, and rushed away with his prey. One of the adult male chimpanzees seized the kill from the youngster and, in a short space of time, the carcass was torn into pieces by three big males in a positive frenzy of noisy excitement. The adolescent hunter joined the females to beg for scraps.

Over the years we have observed many instances of sophisticated cooperation during hunts, and a good deal of food sharing. We now know that chimpanzees hunt for meat throughout their range in Africa—or at least in all places where they have been studied.

NEW LIGHT ON HUMAN EVOLUTION

Louis sent me to learn about chimpanzees in the wild because he hoped it would give him new insights concerning the behavior of our earliest ancestors, and theirs. He argued that if there were similarities in the behavior of modern chimpanzees and modern humans, those behaviors were possibly part of the repertoire of the apelike creature with hominid characteristics, ancestral to both humans and chimpanzees, that lived about seven million years ago. And, if so, then those same behaviors were probably inherited by prehistoric human beings also.

The observations made at Gombe suggested, for the first time, that prehistoric humans may have hunted for meat and used primitive tools made of leaves and sticks long before the first hammer stones and hand axes were made. I love to imagine those earliest ancestors kissing, embracing, and holding hands, visualize their excitement after a kill had been made, picture them using simple tools to help them with the gathering and preparing of their food.

Louis was in the forefront of this kind of thinking, and his vision paid off: Most textbooks now refer to chimpanzee behavior when speculating on the behavior of our prehistoric ancestors.

Today it is generally accepted that although the earliest humans probably ate some meat, it was unlikely to have played a major role in their diet. Plants would have been a much more important source of food. This is true of almost all the hunter-gatherer peoples whose way of life lasted into the last century. The exceptions are when a group of people has moved into an environment that is, for at least part of each year, hostile to plant growth. This is true for the Inuit and the Alaskan Eskimos, and those people who moved into the arid plains. But whatever our prehistoric ancestors ate or did not eat it is safe to assume that the search for food, and the competition with other prehistoric creatures, played a key role in human evolution. For one thing, it was the fact that their diet was unspecialized that enabled our apelike forebears to move out from the forests where, it is assumed, they originated.

Early humans shared the African savanna with many formidable creatures, including giant, gorilla-sized baboons. There was possibly intense competition between them, just as there is today between chimpanzees and baboons at Gombe, where the two species feed on many of the same foods. One such is a tennis-ball-sized Strychnos fruit with a very hard shell. The baboons are easily able to break these open with their strong teeth and jaws, but the chimpanzees cannot. However, the chimpanzees have learned to crack the fruit against a rock to get at the flesh. In West Africa chimpanzees have even developed a hammer-and-anvil technique, cracking open hard-shelled nuts by placing them on rock or root "anvils" and pounding them with rocks or clubs. This innovation gives them access to a rich food supply that is safely out of reach of most creatures. It seems reasonable to suppose that prehistoric hominids also used rocks not only as weapons, but to crack open hard-shelled fruits and nuts.

At Gombe both chimpanzees and baboons love to feast on termites. The baboons—like other monkeys, birds, and so on—must wait until the worker termites open up the nests to enable the fertile princes and princesses to fly off to form new colonies, at which time the eager insect lovers grab as many as possible of the large, succulent flying insects. So do the chimpanzees. But through the skillful use of tools, chimpanzees, as we have seen, can feast on termites even when they are not flying. This opens up a rich source of food when it is not available to baboons and most other competitors.

Amazingly, chimpanzees often steal meat from baboons, usually a bushbuck fawn that the baboons came across during foraging. And this, despite the fact that the male baboon has truly formidable canines, similar to those of a leopard, almost twice the size of those of a male chimpanzee. Even more amazing, the thief is sometimes a female chimpanzee whose teeth are even smaller. It is, I think, because the chimpanzees adopt an intimidating upright stance and charge their adversary, often brandishing a big stick, sometimes throwing rocks, while uttering spine-chilling yells. Given this scenario, we can easily imagine how early humans managed to hold their own against a variety of formidable competitors. Then, as their brains increased in complexity, they would have gradually developed more sophisticated tools and weapons, and eventually gained the upper hand in the savage prehistoric world.

Those prehistoric people also may have learned by watching other animals: Certainly their rock paintings reveal that they were keen observers of the wildlife around them. Perhaps they first thought of coating the tips of their arrows and spears with poison after watching the death throes of the victims of snakes or spiders. And maybe the first clay pot was made by some observant human who had watched the extraordinary skill of the potter wasp as she fashioned mud, chewed in her mouth, into a perfect, globe-shaped clay chamber in which to make a nest for her young.

FIRE—THE ORIGINS OF COOKING

It has been suggested that the introduction of cooked food may have been a major force in shaping human evolution. This theory has been propounded by anthropologists Richard Wrangham (who once studied chimpanzee feeding behavior at Gombe), David Pilbeam, and other Harvard scientists on the team. Charles Darwin himself, as Wrangham reminds us, wrote that cooking provides a means "by which hard and stringy roots can be rendered digestible, and poisonous roots or herbs innocuous." It is possible, too, to extract more calories from some foods when cooked. Wrangham suggests that cooking played a major role in the development of smaller jaws and teeth, and reduction in gut and rib cage size, and that more readily digested food would have provided the increased energy needed for fueling a larger brain.

It is easy to see how a taste for cooked foods could have developed in early humans. Both baboons and chimpanzees sometimes forage in the blackened ground after a bush fire has swept through. It seems they like the taste of singed insects and certain plant foods. And they almost certainly find the occasional dead animal, killed and perhaps partly cooked by the flames. Bush fires in the dry season are often started by lightning, and perhaps, as the human brain became more complex, early humans nurtured the flames for cooking fires. Even a captive mongoose I once knew preferred his meat cooked, and would take pieces of raw steak and push them close to the electric fire.

THE DAWN OF HUMAN CULTURE

Our study of chimpanzees throws light on the dawn of human cultures. For chimpanzees, no longer imprisoned within the cage of instinct, are able to pass information from one generation to the next through observation, imitation, and practice. Sometimes an individual acquires a new behavior by capitalizing on a chance experience, sometimes by watching, then copying, another. And these behaviors can then, in turn, be acquired by others in the group. And although it may be easier for them to learn new behaviors during infancy when the brain is at its most plastic, they can continue to acquire new skills throughout life—unless they live long enough to become senile!

In all the places where wild chimpanzees have been studied there is strong evidence for cultural behavior. Mahale National Park is situated on the shores of Lake Tanganyika about one hundred kilometers south of Gombe. Many of the same species of plants and trees are found in both places. Yet often that which is eagerly eaten by the Gombe chimpanzees is ignored by those in Mahale, and vice versa. In Gombe I have seen older family members "protecting" infants, swatting away foods that are not part of the normal diet of the community—though they may be eaten elsewhere.

Even when the same kind of plant is eaten by chimpanzees in different areas, it may be prepared or collected differently. At Gombe chimpanzees feast on the fruits, the pith, the dry male flower cluster, and the dead wood of the oil nut palm. In the Ivory Coast the chimpanzees eat only the pith. In Guinea the chimpanzees use rocks to crack open the very hard pits and eat the kernels. And in Mahale the chimpanzees ignore the oil palm altogether. Driver ants are captured at Gombe with long peeled sticks that are thrust down into an opened nest. When the stick is pulled out, swarming with viciously biting ants, the chimpanzee sweeps the stick through one hand, then rapidly crunches the great handful of insects. In the Ivory Coast a chimpanzee pokes a short stick into a marching column of the ants, quickly withdraws it when one or two insects climb on, and picks them off with his or her lips. There are many such examples of cultural differences in the behavior of wild chimpanzees.

Thus, chimpanzees have clearly started out along the path of cultural evolution—a path along which we humans have traveled so far in such a relatively short time. A path that has led to the fascinating variations in the foods eaten in different human cultures and the thousands of ways we have discovered to prepare them for the table.

Chapter 2 A Celebration of Cultures

"Food to a large extent is what holds a society together and eating is closely linked to deep spiritual experiences."

—PETER FARB AND GEORGE ARMELAGOS, *CONSUMING PASSIONS: THE ANTHROPOLOGY OF EATING*

Given the diversity of habitats and the diversity of human cultures, it is not surprising, I suppose, to find how many different things are eaten by different people in different parts of the world. In fact, almost anything that is edible—able to be ingested and digested—by human primates is somewhere or other incorporated into the diet. And things considered revolting or even unclean by some peoples are the delicacies of others; or, as the saying goes: One man's meat is another man's poison.

Our palates are shaped, initially, by the culture, the family, the times, in which we grow up. The foods we ate in childhood cause us either to avoid certain dishes—when we were forced to eat them against our will—or to love them because of warm memories. I grew up in Bournemouth, England, during World War II and this left me with a love of tinned peaches and pineapples. When the sirens woke us at night with their loud warning of enemy bombers approaching, we all had to crawl into our bomb shelter—a seven-foot-by-seven-foot, five-foot-high steel mesh cage with a solid steel roof that was issued to families with children. Into this space we had to squash six adults and two children. A regulation amount of food and water was kept there as well—in case we were trapped by rubble from a bomb explosion—and this included a few tins of peaches and pineapples from the food packages sent by generous strangers in Australia. It got pretty claustrophobic in that tiny space, and if we were forced to stay there more than two hours before the All Clear sounded, we were allowed to open one of those tins of peaches or pineapples—a couple of chunks each. My mouth still waters at the thought!

During the war, almost the only eggs we got were dried eggs that came in the care packages from the same generous strangers from Down Under. When the war ended and real eggs reappeared, I always insisted that they be cooked really well—the runny yolk and slimy white of an underdone egg made me sick to my stomach. I was told that if, in a friend's house, I was served a soft-boiled egg, or a sunny-side-up egg, I MUST eat it—it was considered the height of bad manners to reject food served to you as a guest. To this day I feel sick when I even *see* a runny egg—so when my son was small I forced myself to feed him both well-done and underdone eggs so that he should never have to suffer that misery himself. My mother was chronically allergic to shellfish so we never had them in the house when I was growing up. The very idea of eating that luxury of the elite—live oysters—revolts me, yet people will pay a lot of money for the poor things. I did not grow up as part of an elite society, and certainly no one I knew ate oysters when I was growing up in wartime England. Nor do I think I could eat—except under extreme duress—the fat, white grub of an African tree beetle. Yet for many growing up in the forest these are true delicacies, eaten alive and wriggling.

I learned a good deal about food and eating when I worked as a waitress to earn my fare to Africa, way back in

1956. It was a sedate hotel in the seaside town of Bournemouth, on the south coast of England. People came to stay for a week's holiday, so we only got tipped after the full seven days. Very different from a restaurant where people go in for just one meal. Because, in my hotel, the meals were set—I don't even remember there being a choice for the main meal, though I suppose there may have been—many of the guests solemnly ate their way through everything on offer. Though often, I could sense, it was an effort. But they had paid for it and so they were going to eat it. Mind you, they were small postwar portions, not the greedy heaped servings that people expect today which have led to so much shocking waste in the elite societies around the world.

What has always impressed me is how people have been able to find a use for almost every part of a slaughtered animal. The lining of the stomach is sold as tripe—as in tripe and onions. The guts are "lights"—I can never forget the horrible smell as this was cooked for our childhood cats. The brain is a high-class item—sweetbreads. One of the best accounts I have ever read is in Marjorie Kinnan Rawlings's *The Yearling*. It describes, in meticulous detail, the method of cooking, preserving, and utilizing every single part of the family pig's anatomy.

And two other well-loved books, *A Tree Grows in Brooklyn* by Betty Smith and *Angela's Ashes* by Frank McCourt, along with many other books written about those who grew up in poverty, provide vivid descriptions of the struggles of mothers to make ends meet, sending out their children with one or two precious pennies for a few wilted vegetables, or a bullock's eye, or a couple of bones. And there are countless heartrending descriptions of the rations—a crust of bread, a bowl of evil-smelling liquid that was called soup—that were handed out to the victims of the Nazi regime in Auschwitz and the other concentration camps.

DIFFERENT PLACES, DIFFERENT FOODS

Many countries are known for special dishes, part of their cultural heritage and national identity. Hence the now politically incorrect ways of labeling people of different nationalities: The German is a Kraut, the Frenchman a Frog, the Briton a Rostibiff. In some countries a food is even part of the national identity—in Wales it is a leek, in Tanzania ears of wheat, in New Zealand the kiwi fruit. Even in the face of the rapid spread of McDonald's, Kentucky Fried Chicken, Wendy's, and so forth, we are not yet in danger of losing our national food identities: tourists still seek—and find—the food of the people.

Italy, of course, is famous for its pasta, perhaps mostly for spaghetti. (And I wonder if anyone remembers the documentary that appeared on British TV years ago showing Italian peasant women during the "spaghetti harvest"? They were carefully plucking the long strands of full-length pasta from low bushes. It was April Fool's Day!) England is famous for roast beef, roast potatoes, and Yorkshire pudding (crispy batter)—and, of course, fish and chips. And, in postwar years, I was often treated to a "pub lunch" or ploughman's lunch, that consists of beer, crusty bread, and cheese with pickled onion. Cornwall is famous for Cornish pasties, Devon for its teas and scones, thick Devonshire cream and strawberry jam. And Scotland is known for shortbread, black sausages, and haggis.

Germany brings to mind *Apfel strudel,* and sauerkraut with mashed potatoes. I was familiar with Hungarian goulash long before I went to Hungary. France conjures up a host of delicious foods, but also frog's legs and snails—which we must, of course, call escargots. And Holland is famous for its smoked eel pancakes, and plovers' eggs that can be collected during a two-week period only, at the very start of the breeding season. The hen then lays two more eggs, which cannot be touched.

Jewish tradition is rich in its cultural embrace of food: lox and bagels, matzo balls, kreplach, kugel, and knish, and the strict disciplines of kosher cooking—all these and more are identified with the Jewish culinary heritage. The origin of matzo can be traced back to the flight from Egypt—they left in such a hurry that they grabbed their unleavened bread from the ovens and took it with them. Later, when they were hungry on the journey, they ate the loaves of flat bread, now called matzo.

Because, for hundreds of years, Jews have established communities in so many places around the world, their cherished traditions were often influenced over time by those of their non-Jewish neighbors—in Morocco couscous was adapted into Jewish recipes, as was borscht in Ukraine. Even potatoes, the staple in latkes, the potato pancakes long associated with Ashkenazic cuisine, weren't widely available in Eastern Europe until the eighteenth century.

In Jewish and Muslim law, pigs are unclean, and must not be eaten in any form. It has been suggested that this was a smart ruling because whoever made it knew that pigs are likely to be infected with tapeworm (known as measly) and that undercooked pork would make people sick.

Uganda is famous for bananas—of which there are so many varieties—served with a delicious peanut sauce. In parts of west and central Africa almost any kind of wild animal is considered suitable as food. When my son Grub was in Sierra Leone, he found a bat's wing in his soup. And President Julius Nyerere of Tanzania was horrified, on a visit to Zaire (now the Democratic Republic of Congo), to discover the entire hand of an infant chimpanzee on his plate. He was told it was a great delicacy, and he spent the rest of the meal trying to hide it under his salad.

A Masai herdsman in a harsh environment enjoys a bowl of cow's milk mixed with blood not only for the nourishment it provides but because it was such a treat when he was growing up. Once my first husband, Hugo, and I were offered some by two Masai warriors on the Serengeti. It was a horrifying moment for me—although I am not lactose-intolerant, I have always hated drinking milk, even as a baby. And not only was this mixed with blood, but I'd been told that the gourd was always washed out with cow urine. I forced myself to take a sip—but it was mostly pretense, just touching the gourd with my lips and making a swallowing movement accompanied by a delighted smile!

In India, people seem to have a deeper sense of connection with food than in most Western countries. Food plays a significant role throughout Indian culture and is always eaten in joy, never sorrow—in honor of marriages, births, promotions, engagements, anniversaries, a new job, even the purchase of a new car or a home. Before breaking the good news to friends and family, however, Indians pass out sweet candies to help prevent people's tongues from getting sour with envy.

The Western world, of course, thinks of India as the land of curries. So many, all with exotic and mouthwatering names. When I first got out to Kenya, home to countless Asian Indians, I learned to eat and love very hot curries. I remember going out to lunch with a group of young men. They were all sweating profusely. "Everyone sweats unless their bowels aren't working properly," announced one of them—then, noticing that I was not even perspiring, blushed brilliant red. (My bowels were working perfectly, thank you!)

China, of course, is famous for "Chinese food," though the food eaten by the ordinary Chinese people bears little resemblance to that served in Chinese restaurants in the West, as well as in expensive restaurants that cater to visitors in modern China. I love eating this food—in China and as I travel around America during my endless lecture tours.

Japan feeds its people a wide variety of noodles, and serves up seaweed in a whole variety of delicious dishes. But most famous are the sushi bars, and fish of all sorts. Indeed,

Japan is emptying the seas of fish to satisfy the voracious appetite of her people for fresh fish.

America, made up of people who have emigrated from all around the world, shares many of the national dishes of Europe, Asia, and Africa. French Creole influences can be found in the signature foods of Louisiana, from catfish gumbo to jambalaya. In Chicago we find Italian deep-dish pizza. There's the African-American soul food from the South—grits, cornmeal griddle cakes, and other comfort dishes. The delicious blending of Tex-Mex in the Southwest, from nachos to chiles rellenos. In the Northwest you find the traditions and seasonings of Asian immigrants melding with the traditional Native American reverence for seafood. And, in the cattle-raising states, beef, cooked in every way it is possible to cook beef, served in huge amounts on huge plates—the largest servings of all, of course, in Texas.

Meals on the Road

My life since 1986 has been a blur of lectures, airplanes, and hotels. And wonderful people. Though I seldom remember the actual food we ate, I have fond memories of meals with friends, all over the world. The elegant restaurants in Japan, in small private rooms, sitting on the floor on little cushions set on rush mats, served by delightful geisha wearing the beautiful kimonos of yesterday's Japan. The exquisite lacquered or porcelain dishes set out, one following another throughout the meal, each with its dainty offering of some delicious and, to our Western taste, exotically flavored morsels. And an endless supply of hot sake, drunk from tiny porcelain cups, each holding a mouthful. I remember formal banquets in

Taiwan, served with the same reverence by delicate girls in traditional dress. There it seems that the delicious courses will never come to an end, and I have learned to eat only the tiniest amount of each. You do not hurry through these meals but savor each mouthful, which is how it should be if we honor the food that nourishes us.

Some of my fondest memories are of meals with friends gathered in my hotel room. More private and quiet than a restaurant. One hotel, the Roger Smith, is really special, my home away from home in New York City. The owners, James and Sue II Knowles, give me a room (as a donation to JGI), almost always a suite—which is perfect for these evening gatherings of FOJs (Friends of Jane!). We sit around on the floor with Chinese or Indian take-away, manipulating chopsticks in candlelight, catching up on each other's lives, discussing the world's problems, drinking wine and laughing.

CHEERS! SALUTE! SKOAL!

Many countries are associated with different kinds of beverages, alcoholic or otherwise. The British drink draught bitters, gin and tonic—and tea. The Scots, of course, have a monopoly on Scotch whiskey—all those glorious brands of single malt. In the U.K. it is only known as whiskey: We don't use the "Scotch" that in America differentiates it from bourbon. And only Irish whiskey can be used to make that marvelous Irish coffee. Germans are well known for drinking beer in Biergartens. Russians drink vodka at the slightest provocation: It is not unusual, I am told, to be toasted at breakfast with a shot in a small glass, frosted with ice. I was introduced to a kind of vodka in which red chilies had been steeped—an acquired taste, I was told, by my dinner host on the last day of a conference in Moscow. It comes from Ukraine, and the Ukrainian at the table was delighted when I drained several glasses with him—after some wild dancing! The various European countries have their schnapps, Poirre William, aquavit (my favorite), and so on.

France will always be associated with a huge range of exquisite wines. Germany, Italy, and Spain are also known for their fine wines, but now that other countries are producing and exporting some fabulous wines there is a lot of competition. California in the United States, South Africa, Australia, Chile, Argentina, Romania, and Bulgaria, to mention but a few. England has not got much beyond homemade wines, sometimes concocted from dandelions and other plants. My grandmother used to make elderberry wine each Christmas, and when I went to stay with one childhood friend I could never have a bath because her husband always had wine fermenting in the tub!

Japan is famous for its rice wine or sake. I was quite revolted when I first drank it, and it took me a couple of visits to acquire the taste. Now I truly appreciate hot good-quality sake. In fancy restaurants they serve it in wooden boxes!

Throughout Africa inebriating drinks are made by fermenting a variety of ingredients. When I first arrived in Africa to stay with a school friend, the Kikuyu grooms used to make *pombe* by fermenting some kind of grain. This was strictly forbidden, so they hid the bottles in the stables, in the hay. This generated heat, so that the bottles sometimes exploded, sounding like gunfire. I always worried lest the horses be hurt by splinters of glass, but I never heard of any problem.

In many parts of west and central Africa palm wine is the drink of choice. When I visited villages around our chimpanzee sanctuary in Congo-Brazzaville I was invited, along with the chief and other dignitaries, to scatter a few drops of palm wine on the ground before drinking. This was an offering to Mother Earth, to thank her for her bounty. I have shared locally made and somewhat intoxicating drinks, served in wooden bowls, with a group of Achwa Indios with painted faces in a very remote village in the heart of the Ecuadorian rain forest, and in Taiwan with a highland tribe of indigenous people that is trying to rediscover its lost culture.

India, China, and Japan are best known for their infinite varieties of tea. In Japan the ritual of the tea ceremony acquires a spiritual significance. I remember, on one visit, crawling through a tiny opening into an inner sanctum (you could not enter with a sword, and everyone went on hands and knees to show that all men are equal), where I was served bitter bright green tea, the consistency of pea soup, by none other than the Grand Master himself.

Africa and South and Central America are famous for the most exquisite varieties of Robusta coffee, which is consumed by the elite in more and more countries. Turkish coffee is thick and bitter, and drunk in tiny handle-less cups. And there is some debate whether it was Italy or France that first introduced the espresso.

FOOD ON THE STREETS

There are still street vendors in most countries, offering a whole variety of local foods, sitting at little stands, squatting on the pavements, selling from loaded bicycles. Hotdogs; ice creams; little pieces of meat, hot from a charcoal fire, skewered on wooden sticks; fruits of every description cut up and offered in ready-to-eat portions. Young men holding a stack of five- or six-minute cups that they jingle up and down to create the distinctive sound that advertises the hot coffee they carry in jugs. The roast chestnuts and sweet, very hot mulled wine that is so welcome on freezing days at the German, Austrian, Hungarian, and other European

Christmas markets. And, of course, there are markets, where one can find local foods and crafts, everywhere in the world.

The Clay Eaters

Chimpanzees break off and eat little bits of termite clay almost every day. Pregnant women living in the villages around Gombe buy small fragments of the clay from the market. And, I found, so do women in other parts of Africa. Now, to my delight, I read an article that describes how "clay-dirt," a fine-grained subsoil, was eaten by generations of poor whites and blacks in the American Deep South. It was more or less a staple for pregnant women in rural areas.

Anthropologist Dr. Dennis A. Frate studied this strange behavior. Although it has almost died out, he found some who remembered eating clay-dirt as children. One person was Fannie Glass, from Mississippi, who said she really missed having some dirt to eat. "It just always tasted so good to me," she said. "When it's dug from the right place, dirt has a fine, sour taste."

In 1971 more than half the women surveyed in a rural Mississippi county said they had eaten clay-dirt, but by 1984 only one of the ten Frate tracked down still kept up the custom. This was Iris Cornish from Louisiana. She said good dirt is hard to find today—so many of the good places had been covered over with concrete and buildings. She reminisces about eating dirt when visiting her grandmother as a child: "I would sit on her porch with my aunties and a cousin and share some. About a cupful." They used to carry bags or jars of the clay to snack on—sometimes they "baked it to kill the worms" or seasoned it with vinegar and salt. Families used to mail boxesful to relatives who had moved to the North, but still craved the flavor of the clay hills they had left.

Geophagy—earth eating—is known from Roman times, when medicinal tablets were made from soil and goat's blood. And in the nineteenth century, Germans used to spread it on their bread.

Recently, samples from three areas were analyzed. A fine light soil from China's Hunan province, used as "famine food" in the 1950s, was found to be rich in iron, calcium, magnesium, manganese, and potassium. Soft clay from Stokes County, North Carolina, was rich in iron and iodine, often lacking in the diets of the poor. Red soil from termite mounds in Zambia, used by the local people to soothe upset stomachs, contained kaolinite, a main ingredient in many diarrhea remedies in Western medicine.

FINGERS, FORKS, AND CHOPSTICKS

The first tools of choice in the evolution of human feeding behavior were, of course, the fingers. And millions of people around the world still use this method of transferring food to the mouth. In countless homes in Tanzania, both African and Indian, I have watched with admiration the graceful movement of the fingers and thumb as people gather up rice and vegetables, dip them in sauce, and eat. I can do the same, but I know I am nowhere near as proficient, as elegant. In the West we are seldom allowed to use our fingers—how quickly small children, eating in the way that makes the most sense, are taught to exchange fingers for spoon and fork. We can, though, use fingers to eat artichokes—and in Britain we are told that we can pick up chicken bones in the time-honored way, as they did in Merrie Englande, even in the presence of the Queen!

Mostly, though, custom requires us to use knives and forks and spoons—and the array of cutlery laid out for each guest at an elegant dinner party or banquet can be utterly intimidating. Indeed, much has been written about the dismay experienced by those new to the ways of society when confronted, for the first time, by the genteel dining table. Which knife and fork, for heaven's sake, do we use for which course? The bewildering sequence, from soup, to aperitif, to fish, to entrée, to dessert, to savory, and, finally, cheese and coffee. Etiquette is everything. Imagine this, then: An elderly gentleman, dressed impeccably in a velvet smoking jacket, is seated, with the appropriate flourishes, at a table in a very exclusive restaurant. He then proceeds to order—and solemnly eat—an entire five-course meal in the wrong order! Starting with the brandy and cigar and ending with soup polished off with sherry! By this time everyone present is taking surreptitious peeks—or openly staring. I suppose he did it for a wager!

Then there is the art of chopsticks. The fine, tapered variety favored by the Japanese, that enable you to pick up individual grains of rice with great speed and dexterity, to the blunt variety favored by the Chinese and many other Asian cultures. I was lucky: My father was in Hong Kong at some point during World War II, and he brought back, for me and my sister, a pair of ivory chopsticks. He showed us how to use them, and I zealously used my newfound skill at almost every meal for the next couple of years, however inappropriately. When I was first invited to give a lecture in Japan, back in 1984, Dr. Junichero Itani (Jun) was really impressed, and he proudly showed off the skills of his young English friend. I think, though, he was more impressed by my ability to slurp my noodle soup. Slurping is considered the height of bad manners in England, and even when told that it is really polite in Japanese society, most Brits, Jun said, could not overcome their inhibitions. I must admit, I have never managed the satisfied burp that is so polite in many Asian and African societies.

CELEBRATIONS AND FEASTS

All over the world major events are celebrated with feasting, the consumption of large amounts of food and, sometimes, alcohol as well. Feasts among the wealthy during the days of the Roman Empire were sumptuous and elaborate. To celebrate his victory over Pompey, for example, Julius Caesar gave a banquet feast that lasted for two days with 150,000 guests, seated at 22,000 tables! A typical Roman feast took gluttony to a whole new level. The meals were divided into seven courses, starting with hors d'oeuvres, then three entrées, two roasts, and dessert. Romans loved feasting so much that they would often make themselves vomit so that they could then continue gorging on food for five-hour stretches. The French have been known to do the same during special celebrations.

Scotland celebrates New Year's Eve in a big way. After leaping to their feet to sing "Auld Lang Syne" comes the toast to the new year, and as the last chime of twelve o'clock fades everyone raises their glasses, drinks, and then hurls them over their shoulders to splinter on the ground. The first time I went to a birthday party in the home of some of my Greek friends I was horrified when, as the wine flowed ever more freely and the dancing livened, the guests started to hurl plates to the ground, smashing one after the other, while continuing to dance wildly—in bare feet!

In ancient Scandinavia there was a great yule feast every winter, in honor of the god Frey, when flesh of a boar was eaten. It was brought into the banqueting hall with great ceremony and reverence, crowned with garlands of laurel and rosemary. The head of the family laid his hand on the dish, now dubbed the Boar of Atonement, and swore he would be faithful to his family and fulfill all his obligations. Meat could only be carved by a man of unblemished reputation and tried courage, for the boar's head was a sacred emblem, supposed to inspire all with fear. When married couples wished to live in greater harmony they invoked Frey's name, and when they succeeded they were publicly rewarded with a gift of a piece of boar's flesh.

Some Memorable Meals

Every summer during the halcyon years of my childhood, our mother's best friend, Daphne, with her daughters, Sally and Susie, came to stay with us in Bournemouth. Sally and I were about ten years old (Sue and my sister Judy were four years younger) when we planned the first of several midnight feasts. There were strict rules: It had to start at exactly midnight; we had to creep undetected out of the house into the garden; and the food had to be that which we had managed to save during the previous day or two. Once out in the garden we went to our little camp, screened by thick rhododendron bushes, and struggled to light the small fire we had prepared. We always picked moonlit nights, and limited the use of our one flashlight.

The food itself was mostly all but inedible. Pieces of old toast—a bit like chewing leather—squashed slices of cake—all smuggled out in paper bags (no plastic in those days), we produced and pooled our little scraps of contraband from the old tin trunk in which we had stashed it during the day. The best part was cocoa that we made with water boiled over our little fire. We had premixed the cocoa powder with a little milk and sugar, so all we had to do was pour the water into the tin mugs that we used in the garden.

I wonder whether the grown-ups knew about these feasts? I think they must have but they never let on, and later I forgot to ask. It is too late now.

I vividly remember the time my mother, Vanne, left Gombe just before Christmas in 1960. and *I knew* that she and Dominic (our cook) had spent hours planning my Christmas meal. I went up into the mountains as usual on Christmas Day. I never took food with me, and after a day climbing about through the forest I was quite hungry as I set off down the mountain. I looked forward to sitting beside my little campfire and opening the few presents Mum had left, and the letters I had saved up as a treat. And I wondered what delicious food had been planned.

I arrived at my camp just as night fell and I immediately sensed that something was wrong, for the camp was in darkness. No fire had been lit. No hurricane lamp glowed from the tent entrance. I put my haversack down, lit the lamp, and set a match to the fire. Only then did I see the table, laid for my Christmas dinner. A plate, with knife and fork set crookedly, a tin of spam beside a tin opener, and a mug. That was it!

I couldn't find Dominic. It turned out that some friend had arrived with two buckets full of the local *pombe*—the very alcoholic brew made from bananas—and Dominic had spent the afternoon drinking. He was snoring in his tent, out for the count!

I washed, found a loaf of bread and a couple of tomatoes from the tin trunk (where everything was kept to be safe from baboons), and drew my chair close to the fire. And as I looked at my plate I started to laugh. I couldn't stop—I had tears pouring down my cheeks. And I was so glad I didn't care about eating! (I had Vanne's carefully planned meal the next evening, the table laid with flowers, and a very contrite cook—he was still hung over—bringing me the food.)

I also remember vividly a picnic that was prepared by Tom Mangelsen, one of North America's best wildlife photographers. He lives near Jackson Hole, Wyoming, and I was there to give some lectures. When I told him I had one of my rare free days he offered to take me for my first visit to Yellowstone National Park.

When we found a place to eat—a grassy clearing overlooking the water—Tom got out the lunch he had prepared. What a picnic! He had even managed to find a proper wicker picnic basket. There was crisp salad, tomatoes, avocados, crusty bread, several varieties of cheese, peaches—and white wine. All this spread out on the grass on a check tablecloth. Throughout the meal we had the company of a very large and very bold herring gull who clearly believed that he should share our food and came closer and closer, fixing us with accusing eyes. There were probably bears watching us as well.

FOOD AND RELIGIOUS TRADITIONS

From "the land of milk and honey," as Israel was described in the Bible, comes a cornucopia of ancient food rituals unique to the Jewish faith. On the first night of Rosh Hashanah, honey is used as a sweetener for challah bread, and as a dip for apples, with prayers asking God for a sweet year ahead. On the second night a fruit that has yet to be eaten for the season, often a pomegranate, is consumed as a show of appreciation both for the earth's bounty and for being alive to partake of the feast. In the Sephardic tradition pomegranates were said to bear 613 seeds, one for every mitzvah, or commandment, that Jews are historically bound to obey.

The conjoining of food and faith is also found worldwide in many Christian ceremonies. Most well known is the ritual of Holy Communion that symbolizes, for Christians everywhere, the Last Supper, when the disciples sat around a table with Christ, who broke bread and handed it to each of the twelve, saying, "This is my body which is broken for you," then offered each of them wine, "This is my blood which is shed for you," then bidding them, "Do ye this in remembrance of me." Bread and wine were a staple in the diet of commoners in ancient Palestine and thus widely available for use in rituals. Bread and wine were thus transformed as the host in the sacrament of the Eucharist, symbolically representing—or "hosting"—the body and blood of Christ. In many parts of the world, bread and wine are replaced by other local fare to create the host wafers, such as yams and honey in Central Africa.

The Muslim fast of Ramadan is one month long, during which anyone over twelve years old is not supposed to eat or drink anything between sunrise and sunset. During the daylight hours extra time is spent praying, worshipping, and contemplating the Koran. In Tanzania the population is roughly one third Muslim. During Ramadan there is an atmosphere of stoicism during the day, which changes once the sun is set. Then there is a sense of rejoicing in the streets with brightly lit restaurants and delicious smells wafting out into the darkness from all the kitchens. And at the end of Ramadan comes the feast of Eid Mubarak involving three days of feasting and celebrations.

Eighty percent of India's population is Hindu, and food is central to many religious rituals. Symbols of bounty are found throughout Hindu homes as appeals to the gods to bring prosperity into their lives. Abundant foods like bananas, coconuts, mangos, and rice represent the gifts of nature and are commonly used in rites of fertility, such as marriage and the birth of a child.

Food—rather than animals—is "sacrificed" in Hindu

rituals to honor benevolent gods or appease the malevolent ones. Jaggery, a rich sugar often used for medicinal purposes, is often set out to express adoration for the goddess Santoshi. To appease Alakshmi, the goddess of misfortune, Indians put lemons and chiles outside their houses, hoping this will satisfy her malevolent hunger and prevent her from bringing ill will into their homes. Ancient Hindus believed that in death, the soul departs the body, goes to the moon, and returns to earth in the form of rain, where it embodies itself into food, affirming their belief that what has died supports that which lives. Indeed, to repeat the saying from the Hindu Upanishadic literature that started Chapter 1: "In the cosmos there are only eaters and the eaten. Ultimately, all is food."

The Tale of the Giveaway Buffalo

This story was given to me by my friend Shadowhawk, a member of the Washo Indian tribe from the Nevada River basin, and the father of a very special young man.

"A major part of a Washo's life is centered around something called 'Giveaway'—the way of all life. The two-legged, four-legged, the birds of the air, the fish of the sea, all know that to be centered they must participate in the Giveaway. Everything in our universe practices Giveaway in one way or another. Among Native Americans the spirit of giving is very important. We believe that 'without sacrifice there is no real expression of love.' We give away to friends, relatives, and even to people we may have never met before. We give away for many reasons. We give away if we feel good, or are thankful, or if someone is in need. We express thanks, or attempt to spread the good feelings we have, by giving gifts.

"This is a tale of the 'Giveaway Buffalo.' It took place three years ago on the Grand Ronde Indian reservation in northwest Oregon. It was the time of the Spring Equinox, the time of the year when the tribes gather for the most sacred ceremony of the Cheyenne, Lakota, and other plains Indians, the time of the Sundance. This is the beginning of the new year for many Native Americans.

"The Sundance is a twelve-day ceremony of sacrifice, purification, and renewal in which dancers abstain from food and water for four days, dancing from sunrise to sunset while drummers sing ancient prayers and families and friends watch (and dance) from the arbor. Before they dance the dancers and supporters go through four days of purification. The last day of the Dance is the piercing day. The Sundancers are pierced in the chest and tied by leather ropes, which are attached to the upper part of a Sacred Tree. This is done as a sacrifice to the Creator for the healing of a friend or loved one or so that next year the people will have enough to eat.

"After the Sundance is over there is a great Giveaway and a wonderful feast with all manner of good things to eat. One of the special foods at the Sundance is that of Taanka, the buffalo, the sacred one given to the people by the Creator to give life to the Nations. It was their food, their clothing, their medicine, their lodging, their tools, and more. And that brings me back to my story about the Giveaway Buffalo.

"I had been invited to attend a buffalo ceremony on the Grand Ronde. Each year a tribe or member of a tribe donates the meat and food for the Sundance, and the honor had fallen to a Lakotan friend of mine who had a herd of buffalos on the Grand Ronde reservation. A buffalo ceremony is given to ask a sacred one to Giveaway to the people, and if you've never seen one you will be hard-pressed to believe what I'm about to tell you.

"The morning of the ceremony I asked my oldest son, Washo (who was named after our tribe), if he'd like to come with me. He is a young man who loves animals more than anything else and I knew it might be hard for him to watch but I wanted him to witness what was going to happen so that he would learn what the buffalos and other animals knew, that death isn't something to be feared and that it isn't the end but rather the beginning of life.

"He was somewhat apprehensive but wanted to spend some time with me so he agreed to come along. We got to the Grand Ronde valley early in the morning just before sunrise. It was a beautiful Saturday morning with a blue sky and the sun breaking over the rim of the hills. A hawk flew overhead as we arrived at the end of the dirt road leading to a deep valley of green pasture. There was already a large gathering of men, women, and children lined up in rows of twelve across, all facing the east. They were there to sing the honoring song, the death song of thanksgiving for the one who was going to Giveaway.

"Standing in the field with a group of elders was a Lakota medicine man, his name was Sukawaka Luta (Red Horse). The sun had risen over the tops of the mountain now and the people that had gathered in the valley began to sing. The sound hung on the early morning air and echoed back from the hilltops; it was as though the whole valley was full of voices. The sound seemed to come from everywhere. They would sing for a while, then they would turn in unison, twelve across, to another of the four directions and sing again until they had sung in all four directions and then they would begin again. As they sang, the buffalo herds started coming in from all areas of the pasture. As they came they formed a half-circle in front of the elders and Luta. The buffalos just stood there as the people sang.

"Washo warmed himself in the morning sun and watched and listened. He could see that Luta had a prayer stick in his right hand and in the other was a rifle. He knew it was going to be hard for him to watch an animal die, because for him all living things are family and friends, with fear and feelings. He was just eleven years old and I wasn't really sure how this was going to affect him. While I didn't want him to experience the buffalo's death I wanted him to witness the miracle of Giveaway. The people were not hunting the buffalos, they weren't going to take one, or go out and drag one in, they were waiting for one to offer himself as a gift to the people, in the same way that people may offer themselves to the Creator by giving their lives.

"All of a sudden the singing stopped, the valley fell deadly quiet, Luta raised the prayer stick and asked for the one whose turn it was to give away to come forward. A large young bull began to walk toward Luta. As he prayed, the bull walked slowly past the elders and headed straight for the medicine man. Luta handed the prayer stick to one of the elders and placed his right hand out to accept Taanka's sacrifice. When a buffalo comes to offer the Giveaway he will place his head in the hand of the medicine man and then drop his head to die. But just before the young bull reached Luta's hand an older, larger bull came from out of the middle of the herd, and running in front of the young bull he pushed him away and placed his head in Luta's hand. Some of the herd came and circled the young bull as though to hold him back.

"It was quite an amazing sight. There is no greater love than this—that a man (in this case, an animal) lay down his life for his friends. I'm not sure who learned the most that morning, Washo or myself, but I left there more thankful for all those that have given away to me in my lifetime and inspired to give more of myself away than ever before."

Chapter 3 How We Lost Commonsense Farming

"A nation that destroys its soils destroys itself."

—Franklin D. Roosevelt

From the age of fifteen, until I left school at eighteen, I used to spend a part of all my holidays helping at the farm owned by my riding teacher. One of the tasks I remember was "muck spreading"—I would stand on the trailer piled with farmyard manure and toss out forkful after forkful as the tractor was driven up and down a plowed field. As a treat I was sometimes allowed to drive the tractor. We were preparing the field for planting potatoes. I helped with the potato picking, too, when they were ready for harvesting. It was backbreaking work, following along behind the special digger that loosened the soil and brought the potatoes to the surface, and putting them into sacks. The undamaged ones went in one sack, the damaged ones into another. Some of them had worms—fine, they went off to the potato crisps factory, along with those damaged by the digger. They were all pure, organic potatoes, harvested from rich soil, fertilized by the manure from the horses and cows, surrounded by hedgerows full of wildflowers, bees, and butterflies.

All that changed when traditional farming methods and land use gave way to industrial-style agribusiness.

The trouble is, industrial farming typically harms the farmland itself. In the old days farmers rotated their crops and their livestock, and often allowed fields to lie fallow for a year every so often. With this kind of consideration, the land remained fertile for hundreds of years. But when agribusiness steps in and takes over, commonsense land management goes out the window. The big corporations are interested in immediate profit for the short term and show little concern for future generations. More and more land, worldwide, is gradually being killed by libations of chemical fertilizers along with chemical pesticides, herbicides, and fungicides.

CHEMICAL LIFE SUPPORT

The first departure from commonsense farming was after World War II when monocultures were introduced—planting acre after acre with the same crop. Often in the same field year after year. While this may seem a convenient way to make a profit—cutting down on the types of machinery required to plant and harvest, and on the types of chemical supplements—it soon created all sorts of problems. A monoculture is the equivalent of putting all your eggs in one basket. So that if a farmer loses his one and only crop—as a result of disease or adverse weather—he will suffer. In the old days there would have been another kind of crop coming along to help keep his bank balance healthy. So naturally the modern farmer is desperate to keep his one crop alive. He adds chemical fertilizers (many of which contain lead, arsenic, and at times mercury—often introduced through the introduction of sewage sludge fertilizer) to the soil and sprays chemical pesticides on the crop. The insects that prey on that crop begin to build up resistance to the chemicals so that the farmer sprays increasing amounts of pesticides and douses the land with even more fertilizer. And the few remnants of the once flourishing system that dare show themselves are viewed as competitors, labeled as "weeds," and extinguished with chemical herbicides.

Eventually the soil becomes so completely depleted of nutrients that the farm's entire ecosystem is on chemical life support. It is a fearful, stressful form of farming. Perhaps this, at least in part, explains the surge in suicides among farmers since 1998. In the United States and Britain farmers are twice as likely to commit suicide as the rest of the population. Before the advent of chemical farming, the main cause of unnatural death for United States farmers was farm accidents, but today it is at least five times more likely to be suicide. In India in the late 1990s, there were massive crop failures and thousands of farmers committed suicide—many of them swallowed the pesticides they had used their savings to buy, but which had not saved their crops.

It is not just farmers who are vulnerable to the hazards of modern industrial farming. Agribusiness tends to grow only those strains of crops that obtain high yields and maximum market values. Thus the genetic variation that occurs in nature gradually disappears—and it is this variation that can save a particular type of food in times of disease. So when a nation or continent allows agribusinesses to take over countless small farms in favor of a commercially profitable monoculture, at the expense of a diversity of other crops, the entire system becomes vulnerable: An outbreak of disease can suddenly attack billions of plants.

In 1970 almost all the rice crop in Asia was threatened by a virus. This meant that the food supply of hundreds of millions of people was at risk. Scientists searched desperately through gene banks of 47,000 varieties of rice hoping to find one that could resist this particular disease. Eventually they found one—just *one*—growing in a valley in India. So that time the disaster was averted. It is sobering to learn that shortly thereafter that particular valley was flooded for a hydroelectric project. Suppose that had happened before finding the resistant plant . . .

The United Nations board that monitors these things reports that a terrifying number of our food plants are being reduced to just a few varieties that work well for industrial farming: apple, avocado, barley, cabbage, cassava, chickpea, cocoa, coconut, coffee, eggplant, lentil, maize, mango, cantaloupe, okra, onion, pear, pepper, radish, rice, sorghum, soybean, spinach, squash, sugar beet, sugarcane, sweet potato, tomato, wheat, and yam. Wow! The threat is summed up powerfully by Dr. Te Tzu Chang, head of the International Rice Research Institute in the Philippines: "What people call progress—hydroelectric dams, roads, logging, modern agriculture—is putting us on a food supply tightrope. We are losing strains of wild crops and domesticated crops everywhere." The U.S. National Academy of Sciences, commenting on the genetic vulnerability of major crops, had this to say: "America's principal crops are impressively uniform, and impressively vulnerable." And it's not just the U.S.; wherever industrial farming is taking over crops, we find the variety of crops shrinking.

GROWING FOOD WITH POISONS

Ever since World War II, when scientists first figured out that nerve gas used in warfare could be turned on crop-eating insects, the farm industry has became increasingly dependent on the chemical industry. And this has turned out to be an unholy—and a very destructive—alliance. Nature has bestowed all living things with the instinct to survive—adaptation to adversity is the key to evolutionary survival. When chemical pesticides are first introduced into an area, insect predators will quickly be poisoned and die. But gradually, after repeated applications, some insects will build up

Gombe chimpanzee Tubi (to be) feeling a fig to see if it is ripe.
He is sucking the last juices from the large wadge
of peel and seeds in his lower lip.
CREDIT: WILLIAM WALLAUER/ JGI

Old female chimpanzee Flo demonstrates tool using. She is "fishing" for termites. CREDIT: HUGO VAN LAWICK

Infant chimpanzees are fascinated by the performances of their elders and show great concentration as they watch. This is how novel behaviors are passed from one generation to the next through observation, imitation, and practice. CREDIT: WILLIAM WALLAUER/JGI

Topo, an adult male chimpanzee living at a sanctuary in Bend, Oregon, invariably selects organic vegetables and fruits when allowed to choose. Here he eats the organic lettuce and ignores the other.

CREDIT: LESLEY DAY

My picnic with Tom Mangelsen in Yellowstone National Park.
CREDIT: TOM MANGELSEN/IMAGES OF NATURE

Until the advent of the tractor, fields were plowed using horses, oxen, and other domestic animals. The Amish still farm in the old traditional way.
CREDIT: GREG PEASE/GETTY IMAGES

Tearing GM crops from the ground in the UK in protest.
CREDIT: GREENPEACE/MORGAN

Percy Schmeiser, the Canadian farmer who dared challenge the giant, Monsanto. CREDIT: © LILY FILMS

There are still many small farmers who maintain the ancient animal-human contract.
CREDIT: DIANE HALVERSON OF ANIMAL WELFARE INSTITUTE

The life of breeding sows. When they are about to give birth they are moved to farrowing hoops.
CREDIT: CIWF

This duck is being force-fed with a metal or plastic tube shoved down her throat and attached to a pressurized pump.
CREDIT: CIWF

Imagine the frustrations and suffering of the billions of laying hens in "battery farms" like this around the world.
CREDIT: CIWF

resistance. Just as overuse of antibiotics creates antibiotic-resistance in the bacteria that cause sicknesses in animals and humans, heavy doses of pesticides create pesticide-resistance in insects. After more than fifty years of farming with pesticides, there are whole populations of "pest" insects that have evolved to become increasingly impervious to pesticides. The response of the farmer is to spray more often, and with increasingly more toxic pesticides. Nowadays, it's not uncommon for farmers to use three times as many chemicals as they needed forty years ago to kill off the same insects. It's the same situation with using chemicals to ward off marauding weeds, rodents, and diseases: Farmers are using more and more chemicals and finding them less and less effective. Each year, about three million tons of farm chemicals are applied to the surface of this planet.

And all these chemicals, of course, don't just stay on the farm: They escape into the environment. They evaporate into the jet stream and fall in our rain and snowflakes; they are lifted by the wind and drift into our backyards, our playgrounds, our preserved wild lands, and even our organic farms; they sink into the soil and leach into our groundwater, reservoirs, and wells; they find their way into our lakes, rivers, and oceans; and, of course, they can end up in the bodies of animals and people.

What's the collateral damage caused by these chemical assassins? For one thing, it's estimated that only 0.1 percent of applied pesticides reach the target pests, meaning all kinds of innocent bystanders suffer. Sometimes the immune and reproductive systems of honeybees are so compromised by pesticide exposure that they can't produce honey. Agricultural chemicals, combined with industrial and domestic chemicals, that enter the rivers and oceans weaken the immune systems of dolphins, whales, and thousands of other aquatic creatures. They cause birth defects in frogs and other amphibians—such as hind legs that are fused together or extra legs sprouting from their bellies or backs. When orcas are washed up on the shores of British Columbia their bodies are so contaminated with PCBs that they are regarded as hazardous toxic waste. And their calves die from drinking their mothers' toxic milk.

Farm chemicals kill off as many as 67 million American birds each year. I heard the other day that the songbirds that once greeted the spring in Iowa with their joyous chorus have virtually gone from the farming areas. In other words, farming chemicals are destroying our wild flora and fauna. The prophecy of Rachel Carson, in her seminal book, *Silent Spring,* has been fulfilled in many other places.

OUR CHEMICAL LEGACY

And finally, whether it's through the air we breathe, the water we drink, or the food we eat, these renegade farm chemicals also enter human bodies, where some of them stay for years, often throughout life. Because some pesticides have the uncanny ability to mimic hormones, they become concentrated in the food once thought to be the safest for our infants—breast milk. Even a fetus within the mother's womb is contaminated by pesticides, since, like alcohol, drugs, and so on, they are absorbed by the placenta and thus pass through the umbilical cord to the unborn child.

One of the greatest controversies surrounding these farm chemicals is how much exposure is considered safe for humans. There is still a great deal more research that needs to be done, but we certainly know, for sure, that exposure to chemical pesticides is linked to various forms of cancer, as well as Parkinson's disease, miscarriages, and birth defects. We also know that children are especially vulnerable: Until they reach twelve years, their brains and nervous systems are still in the crucial stages of development, so it's especially important for them to avoid those pesticides that directly attack the nervous system.

In 1994 there was a particularly dramatic study on the effects of pesticide exposure, comparing two groups of Mexican children who came from two different towns. The two towns were chosen because the residents in both areas were similar genetically, they ate the same kinds of foods, and they had relatively the same education, economy, and housing. The only noticeable difference between the two groups of children, who were all between the ages of four and five, was that one group lived in the foothills about sixty miles from any agriculture areas, and the other group lived in an agricultural valley where farm and household pesticides were so heavily used that even once common insects, including butterflies, were seldom seen. In the same valley, there were also high levels of multiple pesticides found in the umbilical cord blood of newborns and the breast milk of mothers.

Researchers discovered that the children who lived in the agricultural valley had difficulty performing basic eye-hand coordination tasks, such as dropping raisins in a bottle cap. Pediatricians often measure a child's development of perceptual and motor abilities by how well they can do a simple line drawing of a person. The children in the foothills could draw simple pictures representing people, but the valley children drew lines and designs that did not remotely resemble a person. They also had poorer memory skills and stamina, were more prone to physical aggression and angry outbursts, and were less sociable and creative while playing.

While more information on the long-term effects of pesticides is needed, those of us who keep up-to-date with current research on the effects of various industrial chemicals know for sure that we don't want pesticides in our bodies, our children's bodies, our animals' bodies, or the body of Mother Earth. We don't need more research to tell us that these chemicals are bad for us. We should not tolerate any level of exposure to these dangerous, devilish substances. Someday we shall look back on this dark era of agriculture and shake our heads: How could we have ever believed that it was a good idea to grow our food with poisons?

A HOPEFUL AWAKENING

Once we have woken to the harm caused by agrochemicals, contaminated produce can never again taste really delicious, never be really satisfying, no matter how thoroughly we wash the tomato or how carefully we peel the peach. Something pure and wonderful, something that should be our birthright—receiving wholesome nourishment from the earth—has become tainted and compromised.

And for years, all most of us did was wash and peel and hope that the damage wouldn't be too bad. After all, we could hardly give up our fruits and vegetables—the very foods nutritionists tell us to eat more of. Some people looked far and wide to find food grown without chemicals. Others opted to grow their own. But for the average consumer, there seemed little choice but to resign themselves to a life with pesticides. We'd hear of small, but important victories, such as Meryl Streep testifying before Congress and helping to ban the use of Alar on apples, a chemical considered a carcinogen. But overall, there seemed to be too many farm chemicals and too little power to stop them.

Fortunately, those days are behind us—for now we are being offered alternatives. There really is new hope that we may live to see the end of large-scale farming with chemicals. This harvest for hope is called "organic" food in the U.S. and the U.K., "biofood" in Europe. And this growing trend is changing the course of global agriculture. But before we discuss this hope for our future, we must turn to yet another disturbing element of modern industrial farming—the appearance of genetically modified crops—GM foods.

Chapter 4 Seeds of Discontent

"We simply do not know the long-term consequences for human health and the wider environment. . . . If something does go badly wrong, we will be faced with the problem of clearing up a kind of pollution which is self-perpetuating. I am not convinced that anyone has the first idea of how this could be done."

—CHARLES, PRINCE OF WALES, JUNE 1998, SPEAKING ABOUT GENETICALLY MODIFIED CROPS

Suppose, in a restaurant, you were served a dish of black mashed potato. Chances are you would reject it. In fact, you would be horrified! Even if it was explained to you that this new black color indicated an extra-nutritious kind of potato, you would probably resist, preferring a traditional mound of whitish yellowish potato. But if you were given a mound of mashed potatoes that had been genetically modified to include pesticides in order to prevent insects eating them, you would tuck into it *because you wouldn't know* that this was not the traditional potato of your childhood. And restaurants do not tell you if their food is genetically modified. Not because they are hiding it from you; most of the time they can't tell you because they don't know either.

It was in response to resounding public outrage at the effects of heavy pesticide use in agriculture that agribusiness was forced to consider alternatives. Unfortunately, though, the industry's attempts to move away from pesticide dependency has led to the creation of a technology that could turn out to have equally disastrous consequences for human health and the environment of the planet—the genetically modified organism (GMO).

Genetically engineered products are made by inserting genetic material from one species into the DNA of another. The goal with GM crops is to change the genetic coding so that the crops become resistant to pests and to particular brands of herbicides. For instance, one of the most common genetically engineered crops in the United States is "Bt corn." This corn has been altered to make its own bacterial toxin (*Bt—Bacillus thuringiensis*) that is present in every cell and kills any insects that eat it. The biotech industry tells us that genetically altering crops to contain their own natural pesticide will decrease the need for chemical pesticides and therefore benefit the environment. But we simply do not know the long-term effects of genetically altering foods in this way. All too often, tests to determine the safety of genetically engineered crops are conducted not by objective scientific methods but through studies conducted by the biotechnology manufacturers themselves. Perhaps because of this, it is often concluded that these genetically modified crops can be assumed to be as safe and nutritious as non-GM foods. As a result, millions of acres of GM food has been planted, sold and eaten.

The first biotech crop went to market in 1994. Today, 167 million acres worldwide are planted with genetically modified crops, chiefly corn, cotton, soybeans, and canola, engineered to produce their own insecticides or withstand treatment by herbicides. The United States is the world's top producer of GM foods: 81 percent of its soy, 40 percent of its corn, 73 percent of its canola, and 73 percent of its cotton is genetically engineered, and the technology has also sprung up in many parts of the globe. The rampant proliferation of genetically modified organisms around the world is disturbing to many people—consumers and scientists alike. For one thing there is no guarantee that GM foods will, in the long run, be healthy for the humans who eat them. For another, GM plants pose a very real danger to the environment, for once they are unleashed into the wild, outside the confines of the laboratories where they were grown and tested, there is usually no reliable way to control their spread.

Genetically engineered foods, such as Bt corn, are now being grown and sold throughout North America. Because the U.S. and Canadian governments have been more invested in cooperating with the corporations who make GMOs, rather than the desire of consumers, there is no mandatory labeling of genetically altered products, which means there is absolutely no way to avoid them other than to eat organic foods. But the organic industry may no longer be able to protect consumers either, since GMOs are uncontrollably spreading into the crops of North America and even into Central America.

CAUTION: BIOHAZARDS AHEAD

Imagine how devastating it was for Chuck Walker, the president of Terra Firma, a longtime organic foods manufacturer, to be told that inspectors had found trace amounts of GMOs in his corn chips. Walker immediately collected and destroyed all the packages of "organic" chips, costing his Wisconsin-based company $147,000. Eventually, the damaged corn was traced back to an organic grower in Texas whose crops had been accidentally contaminated by cross-pollination from neighboring farms that were growing Bt corn. So unless we ban Bt corn from being grown, Walker believes its pollen could eventually contaminate every cornfield in the country.

It's not just corn crops that are threatened. Wherever GMOs are allowed to grow there is a risk of infecting neighboring farmlands. Even farmers who follow strict organic farming practices can't protect themselves—because it is impossible to prevent the wind or birds from dropping genetically altered seeds into your field or bees from spreading pollen. The 1999 incident with Terra Firma's corn chips was the first time a U.S. organic food product had to be pulled off the market because of accidental genetic contamination. Since then, numerous organic farms across the country have been contaminated by GMOs. Even the rural, independent farmers of Mexico, whose diversity and purity of corn species are an international treasure, are now experiencing contamination from GM corn grown in the United States.

A LETTER COMES IN THE MAIL . . .

Aside from this threat of rampant contamination, GMOs represent yet another means for corporations to gain global control over the growing and distribution of food. A few multinational corporations now own patents for their GM crops—yes, the DNA codes of biotech plants are now being patented—and they are actually suing the farmers whose crops have been contaminated, claiming patent infringement! So not only has GMO contamination threatened the small farmer's land and livelihood, the companies that own the GMOs claim that it's the farmer who is legally liable!

Percy Schmeiser, a third-generation farmer from Saskatchewan, Canada, had been modestly prospering with his canola crop—a yellow-blossomed plant also known as rapeseed—for fifty years. As successful farmers often do, he had been experimenting with his crops, developing his own naturally vigorous varieties, and each year sowing his own seeds from the previous harvest.

So of course he was aghast to receive a letter from Monsanto, the giant agrochemical company at the forefront of developing genetically modified foods, announcing that they had tested some crops along a roadside ditch on Schmeiser's farm and found Monsanto's genetically engineered canola growing there. According to Schmeiser, one of Monsanto's biotechnology managers must have secretly tested his crops, without permission to enter his fields, because the letter stated that somewhere between 1 to 8 percent of his farm fields were contaminated with their GM canola. The letter went on to accuse Schmeiser of patent infringement and demanded that he pay Monsanto restitution for using its seeds. Schmeiser had never purchased any seeds from Monsanto. The only way their seeds could have infiltrated his crops was from neighboring farms that grew Monsanto's herbicide-resistant Roundup Ready canola.

What makes Schmeiser unusual is not that he received this kind of intimidating letter from Monsanto. Or that he would now have to destroy all his precious heirloom seeds because they were tainted with GMOs. Or that his farm and life's work were now threatened. Scores of farmers across North American have found themselves in the same predicament—suffering from unwanted GMO contamination and then receiving threatening letters from Monsanto. But unlike most of these unfortunate farmers who quietly reach out-of-court settlements, Schmeiser decided to fight back.

With his thinning brown hair, wire frame glasses, and soft-spoken manner, the seventy-two-year-old Percy Schmeiser doesn't look like the type of man to take on a modern-day Goliath. When asked why he was willing to endure years of legal battles and astronomical legal costs, Schmeiser said that his Bavarian grandparents had emigrated from the old country in the 1800s to escape this kind of imperialism.

"They wanted to escape the evil land barons, emperors, and kings that were controlling all the peasants' crops and food," he says. "Now the corporations have become the greedy land barons, emperors, and kings, trying to take control over our food supply. There's nowhere left to flee, we just have to stand up and fight."

After six years of legal battles and $400,000 in legal fees, Percy Schmeiser's case made it to Canada's Supreme Court. Ultimately, Monsanto's patent claim was upheld. But since Percy didn't profit from "infringing" on Monsanto's patent, the Supreme Court ruled that he didn't owe them damages, court costs, or fines. In short, he owed Monsanto absolutely nothing.

Fortunately for Schmeiser, his landmark case was crucial enough to garner worldwide attention—and donations from concerned individuals and foundations. Deborah Koons Garcia, widow of Jerry Garcia, featured him in her excellent documentary *The Future of Food.* But will every farmer be able to get such global support in the face of Monsanto's lawsuits? Will the United States Supreme Court rule in favor of the next small farmer who stands up to Monsanto? What we are looking at now is corporate extortion of North American farmers.

For his courageous, nonviolent efforts to better humankind, Percy Schmeiser received India's Mahatma Gandhi Award. Nowadays when Schmeiser speaks to farmers in India and Bangladesh and other Third World countries, such as in Africa, he reminds them that they have a choice. They can act now to prevent Monsanto and other multinational corporations from growing GM crops in their countries. It's too late for the farmers along the prairies of Saskatchewan who want to grow nongenetically engineered soybeans or canola, says Schmeiser. All their seed supply is now contaminated with GMOs. And since canola is part of the *Brassica* family, it's already cross-pollinating with other crops, such as radishes and cauliflower.

"It's not just in farmers' fields," Schmeiser says. "Monsanto's Roundup Ready canola is now growing in all of our highway medians, parking strips, backyards, schoolyards, and golf courses. We even find it growing in our cemeteries." And because it's genetically coded to be impervious to pests and herbicides, it's almost impossible to kill.

Schmeiser now has a modest organic crop of oats, wheat, and peas. Even though he never used pesticides, he says he once used chemical fertilizer on his canola crop—a choice he now regrets. "I wish I had listened to my wife and mother," he says. "My mother instinctively had a mistrust of farm chemicals and wouldn't allow them on our land. And my wife always insisted on growing all our family food organically. She used to say to me, 'Here you have all these acres of crops that you're selling to other people that have chemicals on them, but at the table we eat organically. This isn't right.' After all we've been through, I know now that I should have been an organic farmer all along."

Even though his legal trials with Monsanto have ended and Percy Schmeiser could choose to retire on his organic farm, he continues to travel the world, battling Goliath. In 2005 he journeyed to Thailand where his public testimony helped convince the government to ban Monsanto's terminator seeds. Just as he began this battle in honor of the ancestors that came before him, he says he continues it to honor those who will follow.

"My wife and I are seventy-two and seventy-three," Schmeiser says. "We don't know how many years we have left and we look at it this way: As grandparents we have to ask ourselves, what kind of legacy do we want to leave our grandchildren? My grandparents and parents left a legacy of land. I don't want to leave a legacy to my children of land, air, and water full of poisons."

LORD OF THE SEEDS

Some of the multinational corporations that are taking over the farmland of North America are also buying up seed companies and literally attempting to patent the world's seeds. This is not science fiction. It's happening now. In January of 2005, Monsanto (now dubbed "Lord of the Seeds") took over Seminis, the world's leading seed company. At this rate, a few multinational corporations are close to gaining control of the world's seed supply.

Even if world hunger could be solved with more food, we haven't seen any solid evidence that genetic engineering is restoring the world's depleted farmlands or increasing food production. Given the speed with which Monsanto, DuPont, Dow, and other multinational corporations are buying up the world's seed supply, some suspect that the main incentive for genetic engineering is financial greed, an attempt by corporations to secure, through patents, control of the world's food supply.

Meanwhile, unsustainable and potentially dangerous agricultural practices are spreading throughout the developing world, where leaders are pressured, often by the World Bank or International Monetary Fund, to cooperate with multinational corporations. Meanwhile the corporations create new markets for their agricultural technology in the developing world and also produce food at low cost (often with slave labor), which they can sell at a huge profit to their clients in the developed world. Thus does colonial exploitation of the developing world continue.

Anyone who has studied world hunger knows that the main reason there are 800 million starving people in the world, and nearly 30,000 children dying of starvation every day, is not due to lack of food per se, but to a variety of different factors such as: political turmoil; poor food distribution; corrupt governments at both regional and central levels; the degradation of soil due to overpopulation and overgrazing; corporate takeover of rural farms with the resulting threat to local farming cultures that were developed to suit local conditions; and finally the hundreds of thousands of peasants who are deprived of their means to live through subsistence farming, and the migration of these peasant farmers to the cities. This last is a tragedy being played out on a grand scale as peasant farmers become increasingly poor and disenfranchised. Often they must leave the land that their forebears farmed for generations and try to get work, for which they are unsuited, in urban centers where there is usually massive unemployment anyway. Thus proud farmers are turned into hungry, often starving, beggars. This is described poignantly in Dominique Lapierre's *City of Joy*.

The real solution to world hunger will only be found when these ills are addressed. We need peace, humanitarian leadership, understanding, compassion, and common sense more than technology, and we are suspicious of the corporate claim that they will eradicate world hunger by converting the planet's farmlands into acres and acres of genetically superior crops that can withstand pests, storms, droughts, and disease.

Thus I was delighted to read an article in the May 2005 edition of the *Ecologist* magazine about an interview with Dr. Tewolde Berhan Gebre Egziabher, the director-general of the Ethiopian Environmental Protection Authority, who is attempting to revolutionize agriculture in his nation. He is vigorously promoting organic farming, encouraging farmers to improve soil quality through composting and crop rotation—especially through the introduction of legumes to increase nitrogen levels—and by terracing fields to reduce soil loss due to runoff. Perhaps his country can avoid the risk of environmental contamination with genetically modified crops and agricultural chemicals—and the resultant payments (bondage?) to Western corporations. In addition, he and other leading Ethiopians are developing local market infrastructures, promoting environmentally sensitive grazing methods, and working to better diversify the nation's market base.

And, Dr. Tewolde said, in his interview with the *Ecologist*, there was no need to buy imported chemical pesticides, since Ethiopia's great diversity and variety of crop species prevents them from having the kind of pest problems that are typical of monocrops planted across large, homogenous farmland. He resoundingly refutes the claim of the industry that GM crops are the solution to hunger in the developing world, insisting that dependence on the technology would "enslave Africa once more—but rather than forcing Africans to grow crops in U.S. soil we would now be forced to grow U.S. companies' crops in Africa's soil."

Dr. Tewolde, we salute you! And what a wonderful example he is setting for other countries in the developing world.

SMALL FARMERS STRIKE BACK

The growing of GMOs raises serious questions about our future harvests. Can farmers' rights to grow conventional or organic crops be protected? How can multinational companies be taken to court for allowing the spread of their GM foods into small neighboring farms? How can farmers maintain the right to save their own seeds? What will happen to the rich diversity of ancient and heirloom crops? Should living organisms, such as seeds, plants, genes, human organs, and animals, be owned and protected by corporate patents? And finally, what will happen to this world if we allow corporations to own life?

Fortunately, we have heroes like Percy Schmeiser all over the world, standing up to the corporations and refusing to let them take over our food supply. Many are so frustrated with the proliferation of GMOs that they are literally ripping these aberrant plants out of the ground. In the U.K., numerous protesters have destroyed GM crops and, when arrested, used the legal defense that their actions were justified because they were "defending a greater public interest"—that is, preventing GMO contamination. There was one landmark case that went to trial in 2004, in which twenty-seven protesters (including Greenpeace executive director Lord Peter Melchett) pulled up a field of GM corn in Norfolk, England. The defense made a convincing argument as to the serious threat of GMO contamination and the judge, in his summary, said that the defendants' actions should be regarded as "legitimate pollution control" and found the protesters not guilty.

In France, there's a group of activists who call themselves the Mowing Brigade and are dedicated to destroying GM crops. The group is led by the famous French farmer Jose Bove, who made headlines in 1999 when he helped demolish a McDonald's near his sheep farm to protest globalization. In 2004 they even organized a Mowing Day where they tore out GM crops in southern France. Another group, the Free Seed Liberation, climbed a barbed wire fence in the middle of the night so they could yank out a crop of genetically modified pineapples near Brisbane, Australia, in 2004. Other protesters in New Zealand took their concerns into the fields and gave battle to newly planted GM corn. In fact, almost everywhere genetically engineered crops are being grown, we find environmentalists, organic farmers, and everyday people desperately tearing them out of the ground. For several years the Brazilian government, hoping to rid itself of GM crops, was paying farmers to yank them out and burn them. But early in 2004, the policy changed and Monsanto was given the go-ahead to grow its GM crops in Brazil.

On June 1, 2005, the first register of GM contamination incidents across the globe was published. Details of all known contamination of food, feed for animals, seed, and wild plants since GM crops were first introduced in 1996 are now available on a Web site launched by GeneWatch UK and Greenpeace. In nine years more than sixty incidents of illegal or unlabeled GM contamination have been documented in twenty-seven countries. The site also provides information on damaging side effects such as the development of herbicide-resistant super-weeds.

The number of incidents across the globe in just nine years throws a sinister light on the information contained in a leaked European Commission internal document, obtained by Friends of the Earth in May 2005. The document was part of Europe's defense in a World Trade Organization dispute with the U.S. over GM crops, and makes it clear that scientists working for the commission have concerns as to the safety of GM foods and crops. They admit that concerns about antibiotic-resistant genes and secondary effects on beneficial insects are "legitimate" and "scientific," and that member states should be able to determine their own levels of protection. Yet since the trade dispute began in 2003 the commission has forced two new GM products onto the market and pressured member states to drop their bans.

Beware of Magic Seeds

Former Peace Corps volunteer Sala Sweet tells the story of working in Walewale, a small arid village in northern Ghana. Sweet worked with a women's group, helping them to create sustainable industry and agriculture in their community. After living in the dry, sun-drenched landscape for a few months Sala realized that it was a perfect environment for growing sunflowers—a hearty crop that could be a viable industry for the women. But whenever she suggested the sunflowers, the village women seemed reluctant and even annoyed by the idea. Finally, after the farmers came to trust Sweet, they told her that an "American seed company" had already suggested a sunflower industry to the Walewale villagers and sold them a supply of special starter seeds at a hefty price.

After the sunflower seeds were harvested, the women found it hard to find a big enough local market for their new crop. So the seed supplier offered to buy their sunflower seeds back from them at a very low price, giving the women almost no profit. The farmers had of course kept some of their harvested seeds to grow for the next crop, hoping to have more success the following harvest. But they quickly discovered that the seed company had given them sterile seeds—also known as "terminator seeds"—which are genetically coded to defy the laws of nature and basically kill their own embryos so they can't reproduce. The seed company then had the gall to offer the villagers another batch of sunflower seeds at an even higher price.

"Now they don't trust Americans and are reluctant to grow sunflowers—a viable crop," Sweet says. "And who can blame them? Look at it this way, some seed company came into their village and tried to take advantage of the women's desire to become self-reliant by selling them genetically altered, sterile seeds that would make the farmers permanently dependent on an American corporation."

What saddens her most is that now someone coming to the Walewale women with good ideas and good intentions is going to be met with resistance and suspicion. Grassroots efforts to help rural communities become self-sufficient are challenging enough, she explains, without this kind of sabotage from the multinationals.

THEY WERE WRONG BEFORE

The multinational corporations that produce genetically engineered foods assure us that their products are safe for human consumption. But there was a time, some sixty-five years ago, when it was believed that the chemical dichloro-diphenyl-trichloroethane (DDT) was harmless for people, animals, and the environment. Yet now the long-term cumulative effect of DDT is known to have wreaked havoc with the environment, causing the near extinction of entire bird populations, including America's bald eagle.

It was the same with CFCs (chlorofluorocarbons). The point is that it is the *long-term cumulative effect* that should worry us when we hear of new products being introduced into the environment. Many GM foods have been created to produce their own pesticides in every cell. What will be the long-term cumulative effect of the massive production of these new toxins? On the environment? On us?

During 2003 some GM corn, produced as animal feed, accidentally got into the human food chain. The worst single contamination incident was when the genetically modified StarLink corn, a variety approved only for animal feed, found its way into the human food chain via taco shells. In order to protect it from boring insects, the corn had been genetically modified to produce its own pesticide. This toxin does not break down in gastric acid—a characteristic shared by many substances that can cause an allergic reaction. Thousands of stores had to withdraw products in seven countries—the United States, Canada, Egypt, Bolivia, Nicaragua, Japan, and South Korea. I was in the U.S. at the time, and a number of people developed alarming allergies—with some going into anaphylactic shock.

Over 80 percent of the soy beans grown in the United States are genetically altered. Like Monsanto's canola, these soy beans are specially engineered to withstand Monsanto's Roundup herbicide. This means farmers can spray Roundup all over their crops and it will kill every living plant except the soy. (Incidentally, Monsanto says its herbicide is as safe as table salt. But researchers have linked the main chemical herbicide in Roundup—glyphosate—to non-Hodgkin's lymphoma.) These GM soy products, which are heavily sprayed with herbicides, are hidden in more than 60 percent of the processed foods in the U.S. You also find them in soybean oil, soy flour, soy lecithin, protein powders, and vitamin E.

And what about all that Bt pesticide in the DNA of the corn supply? At least when a pesticide is sprayed on a plant we can find some comfort with washing and peeling. But what if the plant has been genetically modified so that it contains Bt in every cell, from the root to the fruit? You can't peel or rinse it away. And when those altered Bt cells enter our bodies, what might they do to us?

ANIMALS AND GMOS

Many animals appear to have an instinctive aversion to GMOs. For instance, wild geese avoid eating from fields of genetically altered canola, favoring canola from non-GM fields. *Well Being Journal* ran a story in September of 2003 about a farmer named Bill Lashmet who performed a feeding experiment with his cows. He filled one trough with fifty pounds of genetically modified Bt corn and the other trough with natural shelled corn. He watched as every single one of his cows sniffed the Bt corn, withdrew, and then moved on to the natural corn, which they devoured. American journalist Steven Sprinkel Yankton wrote a fascinating article called "When the Corn Hits the Fan" for the eco-agriculture magazine *ACRES USA* in 1999. According to numerous Corn Belt farmers, hogs won't eat their full rations if GM crops are in the trough. Raccoons frequently raid organic cornfields but won't touch GM corn. He describes one farmer observing a herd of forty deer "mowing down his tofu beans" while across the road there wasn't even one doe gnawing on Monsanto's Roundup Ready soy crop.

Mice and rats don't like genetically engineered foods either. Farmers in Canada and Holland both report that if they store GM and non-GM grains in separate bins, the non-GM bins are full of mice, whereas the GM bins are clear. Lab rats, who usually love tomatoes, refused to eat "FlavrSavr" tomatoes, which were genetically altered to delay their ripening, according to Roger Salquist, a scientist involved in creating the GM tomato. After the rats were force-fed monitored doses of the GM tomatoes, several rats developed stomach lesions, and seven out of the forty rats died within a few weeks. Nonetheless, the FDA gave approval to put the FlavrSavr tomato on the market in the early 1990s with no further testing. Interestingly, it never sold well and has since been taken off the shelves.

The most thorough testing of the effects on animals of eating genetically modified food was carried out at the Rowett Research Institute in the U.K. by a Hungarian scientist, Dr. Arpád Pusztai, working for the British government. His tests were triggered by the creation of a Bt potato (a potato that makes its own bacterial toxin) in the United States.

Pusztai decided to create a GM potato using another natural pesticide—a lectin found in snowdrops. He fed it in large doses to rats, who suffered no apparent ill effects. He then spliced the lectin gene into the DNA of a potato. When he tested his newly created potato on rats he was shocked by what he found. First, his tests showed that the nutritional content of the new potato was different from the non-GM potato parent line from which it had been derived. One crop of the new potatoes contained 20 percent less protein than its own parent line. Puzzled, he continued to analyze the potatoes, and found that even the nutritional content of sibling potatoes, derived from the *same* parent and grown in *identical conditions*, was different. These results suggest that the U.S. Food and Drug Administration policy, based on the *assumption* that GM foods have the same nutrient levels as their parents, is misguided.

But his next tests were even more disturbing. When he fed his new potatoes to rats they suffered weakened immune systems, and the thymus and spleen were also damaged. Some of the rats had smaller, less-developed brains, livers, and testicles. Yet others had enlarged tissues, including pancreases and intestines. These serious effects were apparent within ten days of eating the potatoes. And some persisted after 110 days—the equivalent of about ten years in a human. However, when fed cooked potatoes, the rats remained healthy.

Pusztai was concerned. His tests had been many and meticulous. He had fed rats his GM potatoes, natural potatoes, and potatoes with the same amount of pure lectin as contained in his GM potato. Only when the rats ate the raw GM potato did they suffer the serious negative effects. It seemed to Pusztai that if it was not the lectin that was responsible for the adverse symptoms, it had to be the *process of genetic engineering itself* that had caused the damaged organs and immune dysfunction.

This was unexpected and shocking. Pusztai had already reviewed a number of documents describing tests carried out by other scientists on GM foods that were already on the market. He had been appalled by what he considered the lack of design, superficial nature, and the small number of these tests. He realized that if he had used the same superficial testing as that which had led to the thumbs-up for the Bt potatoes, corn, cottonseed, and soybeans already on the market, his potatoes would also have been approved. They, too, would have been sold to hundreds of thousands of people and could, theoretically, be creating health problems similar to those he had observed in his rats—problems that could take years to appear in humans. Fortunately, we do not eat these foods raw, but the same danger will apply to tomatoes and other GM fruits and vegetables that are not always cooked.

It was at this point that Pusztai discussed his findings on the TV program *World in Action*, during which he said that he himself would never eat GM food again. This, not surprisingly, created a good deal of media interest. Shortly thereafter he was fired, and for a while he was ridiculed and blacklisted in the scientific community.

I heard no more about him for the next couple of years. Then, by chance, I tuned into the BBC World Service when I was in Africa and heard that a group of scientists had come to defend Dr. Pusztai and strongly upheld his integrity as a scientist. This research has now been published in the prestigious British medical journal *The Lancet* ("Research letter: Effect of diets containing genetically modified potatoes expressing Galanthus nivalis lectin on rat small intestines," by Stanley W.B. Ewen and Arpad Pusztai). Of course these new findings are being fiercely debated. The controversy is likely to continue since so much is at stake: the profits of the biotech companies on the one hand; the health of human beings, animals, and the environment on the other.

There are a number of other studies, less rigorous but of great interest. The BBC News, on April 27, 2002, reported that safety tests on a variety of GM maize (corn) had been flawed. At the time of the report the maize, T25, was actually growing in British fields. The corn, apparently, had been tested on chickens. During the test period one group of chickens was fed with the GM product while the other got regular corn. Twice as many of those eating the GM maize had died. Yet T25 had been given market approval. It is because of the known risks and all the uncertainty that some countries have banned the growing and selling of genetically engineered foods. Many residents of these countries are highly suspicious of GMOs and are especially watching American children to see if there are any long-term affects. The children of North America have now become the world's lab animals on whom to study the long-term effects of eating GM products.

WHAT YOU CAN DO

How can we turn the tables, return to a healthy, safe way of farming, in the face of such a daunting and rapidly growing corporate takeover? How can we protect our bodies and the earth from genetically engineered organisms? Fortunately, there are some effective things we can do to help stop the spread of GMOs.

Demand Labeling

The United States is one of the only industrial nations in the world that doesn't demand that genetically altered foods be labeled. U.S. consumers can choose not to buy a packaged food that contains MSG, Red Dye #2, or even salt. But what consumers won't see on a standard label are all the blind GMOs. Without honest labeling, health-conscious consumers have no way of knowing if they are feeding their babies genetically altered formula, or serving their families veggie burgers made from genetically modified grains. Even our fresh produce, such as cabbage, lettuce, potatoes, tomatoes, and eggplants, could all be genetically modified.

The companies that manufacture and patent GMOs know that consumers are wary of their laboratory-altered "Frankenfoods." In fact, most U.S. consumers say they want mandatory labeling of GMOs, because they want to avoid them. It's not surprising, then, that corporations have vehemently fought the labeling of GMOs in the U.S. They know it would be the death knell of the industry. So whenever possible, pressure your political representatives to demand that the presence of genetically altered foods be labeled.

Focus on Grocers

Anti-GM activists also suggest that customers pressure supermarkets to stop carrying hidden GMOs in their store brand products. In the U.S., store brand products account for as much as 40 percent of supermarket sales. So imagine what would happen to the Frankenfood industry if stores stopped buying GMOs for their brand-name products.

One successful campaign began in the U.K., when activists went to supermarkets and filled up carts with packaged foods. When they got to the cash registers, shoppers refused to pay until the store manager gave a personal guarantee that all the items were GM-free. Other activists stuck Biohazard labels on products that were known to contain GMOs.

One by one, the supermarkets gave in. By 1999, most of the U.K.'s major grocers agreed to drop GM foods from their own brands. In the U.S., consumer pressure convinced Whole Foods and Trader Joe's to stop putting any genetically altered foods in their store brand products.

Even if the government isn't responsive to consumers' demands, grocery stores have to be, since we can now take our business elsewhere. If we don't like one grocer's policies, we can find stores that offer a wide selection of non-GMO, organic foods. So wherever you shop, talk with store managers and owners—making your request known and threatening to take your business elsewhere unless they change. If you successfully convince your grocer to carry organic foods, be prepared to follow through on your demands by buying their organic products.

Avoid Eating GMOs

Until we get GMOs labeled, they will remain hidden in our food supply. If you feel inclined to reduce your exposure, the only way you can guarantee that you're not eating genetically altered foods is to buy organic. Sometimes, though, organic isn't available, especially when you are traveling. Here are some other tips for reducing your intake of GMOs.

As much as possible, avoid the top three GM crops: soy, corn, and canola. Be especially cautious with packaged foods, since GMOs from these three crops are used in 70 percent of processed foods in the U.S. Most fast foods also contain a lot of hidden GMOs, since the industry relies on corn syrup for its sweeteners, and soy for its oils and fillers. Sadly, even popular "health foods," such as tofu and soy milk, should be avoided unless they are organic. And remember, over half of all genetically modified crops grown in the world are fed to animals. So be sure to avoid nonorganic animal products as much as possible.

If you visit the True Food Network Web site (listed in the Resources section), you'll find an extensive shopping list of all the brand names of items that contain GM foods and those that don't. It also lists places where GM crops are being tested throughout the world and provides ideas for stopping their spread.

Chapter 5 Animal Factories: Farms of Misery

"The question is not, 'Can they reason?' nor, 'Can they talk?' but rather, 'Can they suffer?'"

—JEREMY BENTHAM

When I was a child and went to stay on my grandmother's farm in Kent it was exciting because there were always so many different animals in the fields and farmyard. Cows grazing, or lying and chewing cud. Two or three cart horses, standing under the shade of a tree—for at that time they were still used around the farm even though most of their duties had been taken over by tractors. Pigs, some suckling piglets in spacious sties, others roaming the fields. Hens and roosters scratching, clucking, and moaning in the farmyard. Chickens, yellow balls of fluff, hastening desperately to peck the ground where their mother calls. Ducks on the duck pond. And a small flock of geese—of which I was respectfully somewhat afraid.

A farmer kept different kinds of animals partly because he knew that this diversity—such as cows, pigs, and birds—worked together in an elegant system that helped his farm thrive. His small herd of cows grazed on a pasture full of fresh grasses, herbs, and clover, with all their rich beta-carotene and nutrients. After some months the farmer moved them to a different pasture, and turned some pigs into the vacated field. Pigs are omnivores. With their strong snouts they can turn up the soil (unless they have rings in their noses) and find all manner of nutritious roots and insects. They even get the goodness out of cow dung, and because their digestive tracts have such a high acid content, they are what is known as "dead-end hosts," killing off any parasites and bacteria that may be present. The pigs also get a variety of immune-boosting minerals from the soil they eat.

A field that has hosted a group of pigs makes a perfect hunting ground for a flock of poultry. The birds peck up worms and other insects in the churned-up soil, at the same time depositing their own high-nitrate manure on the grass, thus ensuring that it will be lush and healthy for the next herd of cows.

The old system of farming, in fact, closely mimicked nature. During the years I spent on the Serengeti plains in Tanzania I watched the migratory herbivores, such as wildebeests and gazelles, move across the grasslands leaving them enriched by their droppings and thus benefiting the other animals, such as the warthogs and the countless species of birds. Of course, the Serengeti ecosystem, like all natural ecosystems, is far richer and more diverse than any farm. For one thing it boasts its complement of carnivores that keep in check the prey animals. Most farmers wage incessant war on predators such as wolves, coyotes, foxes, and birds of prey. This means that the natural prey of the predators increases dramatically. Rabbits, deer, all manner of rodents, and many species of birds are only too happy to add farm-grown produce to their diet, so the farmers have to keep the numbers down themselves. Most of this hunting simply added to the food supply of the local people who enjoyed rabbit pie and venison.

What Will Become of the World's Nomadic Herders?

The nomadic pastoralists, like the Masai of East Africa, are absolutely dependent on their cows (or goats, reindeer, yaks, or camels) for survival.

My late husband, Hugo, and I got to know some Masai when we were working in Ngorongoro and on the Serengeti. They have a love of and deep respect for their cattle, and eat their flesh only on very special occasions, but, in addition to milk, they make use of the cows' blood. This is obtained by plunging the tip of a sharp arrowhead into the jugular and bleeding the animal—if it is done properly the cow appears to feel little discomfort.

Sadly, the days of the nomadic pastoralists are numbered. They are gradually being forced to adapt to a sedentary way of life—sometimes by government decree, often because the vast plains on which their lifestyle depends are shrinking, losing ground to mushrooming human populations, severe droughts, overgrazing as herds grow too big for the remaining habitat, and erosion. This is true in Africa, Mongolia, and Afghanistan, and it is the same for those who herd goats, yaks, or reindeer as well as cattle. Thousands of pastoralists who once roamed the country as free men and women now try, often unsuccessfully, to find work in alien cities, try to adapt to a lifestyle to which they are neither attracted nor suited. Like the Native Americans and the Australian aboriginals, who in some cases lose their grip on life, take to drink, and become beggars. And what happens to the yaks, the goats, the camels—and the cattle?

But I suddenly discovered, in the 1970s, that everything in the farming world had changed. Someone gave me a book by the Australian philosopher Peter Singer. And it was when I read *Animal Liberation* that I first heard about the horrors of "factory farms," the large-scale intensive method of raising more and more animals to supply a growing consumer demand for more and more meat at cheaper and cheaper prices.

Since then I have learned a great deal more about the suffering of billions of farm animals around the world. Clearly, the root of all this anguish is the fact that these farm animals are treated as though they are mere *things*, yet they are living beings capable of suffering pain and fear, knowing contentment, joy, and despair. They surely deserve the right to live in conditions that allow them to express as many natural behaviors as possible. Pigs should be able to root in the earth and their piglets to play, chasing one another and squealing in excitement. Cows need to graze on green grass while their calves gambol in the morning sun. Poultry of all sorts should be able to scratch and peck in the ground and stretch their wings. And all farm animals deserve a bed of straw.

The industrial model of factory farming simply doesn't find it efficient or profitable to consider animals as sentient beings. Instead they are treated as mere machines, turning feed into meat or milk or eggs. As though they have no more feelings or rights than a vending machine.

THE PLIGHT OF POULTRY

Much of our poultry is raised in "battery farms," buildings in which hundreds of cages are stacked one on top of the other. In battery farms with laying hens, a single shed may contain up to 70,000 caged birds. The hens are crammed four or even six together, into small wire cages, so close they cannot stretch their wings. Because they then tend to peck one another, their beaks are often "trimmed" in a painful de-beaking process. And because their claws frequently get caught in the wire mesh on the floor of their cages, they are sometimes trimmed by cutting of the end of the toes so that they cannot grow again.

When the laying hens' egg production starts to decline, they are starved and denied water for several days, and the cycle of light and dark is reversed. This forced molting shocks the birds into losing all their feathers and starting the egg production all over again, though only for a few weeks. After this the exhausted, wasted birds are used for chicken soup. In these egg factories, newly hatched male chicks are considered useless "by-products" and are typically thrown into plastic bags, where they suffocate as more and more little bodies pile up on top of them. Then they are disposed of in garbage bins. Some chicks are ground into animal feed—sometimes while they are still alive.

Broilers or capons—the fowl you buy to roast, or casserole, whose legs and wings and slices of breast you buy in neat packages—are crowded into small sheds where they jostle one another, walk over one another, and trample on those who die.

The short lives of turkeys, pumped with growth hormones until they actually cannot stand, or breed normally, are grotesque. Yet when families gather around the festive table to celebrate Thanksgiving or Christmas, how many will spare a single thought of gratitude for the turkey whose suffering has provided their feast? The turkey spirits hovering there, however, will surely join in the Thanksgiving celebrations—only they will be giving thanks for the death that ended their lives of torture.

Ducks and geese are farmed for their meat in the same kind of intensive factory conditions as chickens. The method by which they are force-fed so that their livers expand to the size that makes the production of pâté de foie gras economically sound, is pure torture. Farmworkers thrust a metal pipe down the duck's or goose's throat as a pump shoots massive amounts of corn feed directly into the poor animal's gullet. In just a matter of weeks, the ducks and geese become grossly overweight and their livers expand up to ten times their normal size. Ducks and geese raised for pâté de foie gras can barely breathe, let alone stand or walk. Many of them suffer from lacerations in the throat, impacted food in the esophagus, and bacterial and fungal growth in their upper digestive tracts.

NOT ALL PIGS LIVE LIKE BABE

In some ways the lot of intensively farmed pigs—or hogs as they are known in the farming world—is the worst of all. For pigs are highly intelligent—at least as intelligent as dogs, sometimes more so. For example, a pig named Hamlet is able to move a cursor (designed for use by a chimpanzee) into different colored boxes on a computer, using his snout; a Jack Russell terrier had not learned to do similar tasks after a year. As youngsters, factory-raised pigs are kept in horribly overcrowded pens, usually on cement or slatted floors. Living in such close proximity, denied outlet for their exuberant energy, they sometimes chew off one another's tails. And so it has become the practice to cut off their tails at birth. In order that they shall put on weight as quickly as possible they are given growth hormones. When they are taken to slaughter, their legs, weak from lack of exercise, sometimes actually break as they try to support their unnaturally heavy bodies. Then they are dragged, squealing in pain. And soon they are squealing in terror. People who have experienced animal slaughterhouses have told me that pigs, very clearly, know what is about to happen, and struggle fiercely to avoid that last walk.

Sows kept for breeding are confined in tiny individual stalls so narrow they cannot turn around. Deprived of all opportunity to express their natural behavior, they bite at anything they can reach. Then they give up, become listless, behave as though "in mourning," head lowered, eyes glazed. When ready to give birth, a sow must endure the torture of the farrowing hoop—a metal cage in which she must lie, unable to stand or turn over, pinned on her side so there can be no danger that she might squash a precious piglet. To a free sow her piglet *is* precious—precious because it is her baby. She will not lie on it. Only in a tiny pen, deprived of any chance to nurture her young in pig fashion—make a nest, lick them, shower them with porcine love—only then may she accidentally crush one. But to the industrial farmer her piglet is "precious" only for its potential monetary value.

Stories of Animal Escape and Rescue

Amazingly, some animals have managed to escape the torture of the slaughterhouse, and when this happens it often makes for headlines. Thousands of people developed a sympathy for pigs as a result of reading E. B. White's *Charlotte's Web* and more recently watching that delightful movie *Babe*. But it is not only in fantasy that pigs escape torment and live happily ever after. In 1996 two pigs, who were subsequently named Sundance and Butch, escaped from an abattoir in Wiltshire, England, swam a river, and managed to elude capture for eight days. The story attracted huge media attention and captured the headlines—and the hearts—of the British people. Indeed, news of the dramatic escape spread around the globe. On the eighth day, a British newspaper paid large sums of money to buy the pigs and sent out a special rescue squad.

There was a desperate hunt that lasted all night and involved the police, a veterinarian, members of the RSPCA, a spaniel, and a lurcher. And despite the pouring rain the media was there, too. At one point there were 150 photographers and TV crews representing all the main British channels and papers, and many from Europe, America, and Japan. Eventually Sundance and Butch were captured and sent to an animal sanctuary where they are still living out their lives in peace. As a result of the publicity, and all the articles that were written about pigs, their intelligence and charm, more people, it is claimed, gave up eating meat. They put two and two together—pork, bacon, ham = pig!

Cinci Freedom, a 1,050-pound Charolais cow, was able to elude capture for two weeks after jumping a six-foot fence to escape a Camp Washington slaughterhouse in Ohio. The first attempt to tranquilize her failed: Workers managed to rope her but she bolted down an embankment and into a backyard, dragging two men with her. She had to be tranquilized again before she could be loaded onto a truck and taken to a sanctuary. This story also captured people's imaginations—the mayor of Cincinnati even decided to offer her the key to the city! She is now living peacefully with another escapee, a cow called Queenie, who successfully escaped a slaughterhouse in Queens, New York. The two became instant friends and are fond of grooming each other and posing together for snapshots.

It was big news in the U.K. when a turkey destined for the table at Christmas fell out of the truck that was carrying him to a slaughterhouse. It seems that he managed to travel for some three miles and arrived, amazingly, outside the gate of a bird sanctuary! (Had he been one of the super-fattened hormone-fed turkeys from an American farm he would likely have struggled just to be able to walk a few yards—and probably would have broken his bones when he fell.) However, apart from looking "a bit disheveled" he was fine. No turkey farm claimed him, so he was taken in by an animal sanctuary, fed corn and other grain, and named Terrence. First he was put with the domestic hens, but he took a fancy to a spectacled owl called Bins. So that is where he was allowed to live.

BEEF: THE HIDDEN STORY

The life of a factory-farmed beef cow is equally miserable. Many are kept in "yards"—crammed into small enclosures, sloshy with mud and feces, or baked hard by the sun, often with no relief from the weather, hot or cold. The ones that are raised on grass and are given freedom on the great cattle ranches probably enjoy at least some periods of their lives. But then comes the roundup, the cruel branding, and the castration. I've read descriptions so vivid that I could feel—and smell—the fear, the burning hair and flesh, the pain. And then, for all of them, comes slaughter.

Mostly they are forced into trucks, or into cattle cars on the railway. Their journeys to the slaughterhouses may take days, during which, although many countries have legislation mandating feeding and watering at specified times, these are typically ignored. A cow who falls—a "downer"—is likely to be trampled to death. If not she will be prodded and hit, then, unable to walk, pulled out despite the pain of a broken leg. And then the killing starts. I shall not describe this here. If anyone wants to know the ugly truth, the number of individuals who manage to avoid the "humane" electric bolt and go through the assembly line fully conscious, it is all written in detail in *The Ten Trusts*.

Is there no legislation outlawing such cruelty? Of course there is—it requires that every cow be rendered senseless before being skinned and dismembered, but in agribusiness today each second represents dollars gained or lost. Inspectors are generally not allowed into areas where they would be able to see violations. Mostly their work consists of checking the emerging dead animals for illegal fecal contamination. Thus the humane regulations are seldom enforced. All of this was documented during an undercover operation in one of the largest slaughterhouses in the U.S. in Washington state, by Gail Eisnitz in *Slaughterhouse: The Shocking Story of Greed, Neglect, and Inhumane Treatment Inside the U.S. Meat Industry.* The raising, transport, and killing of pigs and sheep is just as horrific; the assembly lines where chickens, turkeys, and geese are hung for slaughter just as cruel.

Henry Ford's Assembly Line

It is of more than historical interest to reflect that Henry Ford modeled his assembly line car production after visiting a Chicago slaughterhouse in the early 1900s. He watched the suspended animals, legs shackled and heads downward, on a moving conveyor as they traveled from worker to worker, each of whom performed a step in the slaughtering process. Ford immediately saw that it was a perfect model for the automobile industry, creating an assembly method of building cars.

More than efficient, the slaughtering assembly line offered workers a newly found detachment in the whole messy business of killing animals. Animals were reduced to factory products and the emotionally deadened workers could see themselves as line workers rather than animal killers. Later, the Nazis used the same slaughterhouse model for their mass murders in the concentration camps. The factory-style assembly line became a way for Nazi soldiers to detach from the killing—seeing the victims as "animals," and themselves as workers. Henry Ford, a rampant anti-Semite, not only developed the assembly line method later used in the Holocaust, he openly admired the Nazis' efficiency. Hitler returned the admiration. The German leader considered "Heinrich Ford" a comrade-in-arms and kept a life-sized portrait of the automobile mogul in his office at the Nazi Party headquarters.

There are people who eat venison, believing that it is ethical to eat the meat of an animal that has lived in the wild, and met a clean and honorable death at the hands of a hunter. Indeed. But now we find that deer and other wild animals are being bred for their meat, crammed into small enclosures, subjected to the same kind of factory farm conditions as those that the more traditional farm animals must endure. People eat fish, believing they are eating the flesh of creatures who lived freely in the oceans or rivers, cold-blooded creatures that feel no pain. Yet often the fish are farmed. And certainly they can feel pain. But we shall talk about what goes on in the water in another chapter.

NOT-SO-CONTENTED COWS

What about milk products? What goes on in the dairy industry? Although no one knows for sure, it is believed that cows were first domesticated in southeastern Europe about 8,500 years ago. Since then, dairy products—milk, butter, cheese, and yogurt—have been staples in the diets of millions of people around the world.

Until dairy cows were given hormones to artificially increase their milk yield, even the most productive breeds had udders of reasonable size, and the weight of their milk did not cause them pain when they came in for milking. And in those days calves were often permitted to stay with their mothers for several weeks: Gradually the calf was weaned and the milking of its mother increased, so there was a smooth transition for the cow from feeding her calf to giving her milk to humans. And the weaned calves, even if destined for veal, were often given space to gambol and frolic until their short lives were ended.

How different the life of a dairy cow and her calf is in the modern intensive factory farms of Europe and North America, and other "progressive" countries. On many such "farms" the cows, who were taken from their mothers just days after their birth, never feel grass under their feet. They spend their whole lives tied in long rows in narrow stalls. They stand on cement; they are milked by suction machines. Often they are fed Bovine Growth Hormones so that milk production is dramatically increased—some cows are producing 100 pounds a day—so that even if these overproducers are lucky enough to spend time in fields, they develop huge udders that bulge, bloated and uncomfortable, hindering their movement as they walk to desperately needed milking. Often udders and teats become infected, but there is often no time on these factory farms to deal with such minor (though very painful) ailments. The prophylactic dose of antibiotics in the animal feed is supposed to deal with things like this.

Factory farm cows are usually forced to give birth to a calf every year. Just like human beings, cows have a nine-month gestation period and so this annual birthing schedule is extremely hard on the mothers. They are also artificially reimpregnated while they are still lactating from their previous birth, so their bodies are producing milk during seven months of their nine-month pregnancy. The Bovine Growth Hormone, which forces this heavy milk production, also leads to birth defects in their calves.

VEAL: WHAT YOU'VE HEARD IS TRUE

But even when all goes well at calving, both mother and infant suffer as they are brutally separated. Often I have heard the anguished calling of both cow and calf—it may continue throughout several days if they can hear each other. The calf is desperate for the milk and love of its mother, and she is in torment because she cannot nurture her precious baby. The female calves are raised to replace the worn-out dairy cows in the milking herd. Many of the male calves are raised in crowded yards to become beef, or (the lucky ones) killed at just a few days old to be sold as low-grade meat for cheap food products such as frozen TV dinners. And of course some are used for veal.

Today many people are aware, thanks to animal rights activists, of the tiny twenty-four-inch-wide, bare "veal crates" in which calves destined for "white" veal are imprisoned. They cannot lie in comfort, they cannot even turn around. And to give their flesh that white color demanded by the gourmet diner, for the last few weeks of their pitiful lives they are fed an iron-free diet. The calves are so desperate for the mineral that they try to drink their own urine. At the end, after sixteen to eighteen weeks of suffering, the calves are dragged from their prisons, their legs so weak that they can scarcely walk. Indeed, they often break their legs on their way to slaughter.

THE BIRTH OF FRANKENFOODS

Consistent with thinking of an animal as an assembly line product, scientists are now tinkering and experimenting with the DNA of animals—trying to create individuals who grow faster and produce a quicker profit. One recent genetically altered creation is a super-size beef bull, known as the Belgian Blue. These enormous creatures have 20 percent more muscle (meaning lots more beef to sell) and weigh three quarters of a ton. These poor bulls don't have the bone density to support their own flesh, can barely stand or walk, cannot mate—so the cows must be artificially inseminated, and birth is by cesarean. Scientists have also produced genetically altered fast-growing pigs whose frail legs are so tiny compared to the rest of their bloated bodies that they suffer from painful joints and have difficulty moving. The genetically altered fast-growing chickens are prone to heart disease and their bones are so feeble they break on contact. Genetically altered turkeys are so fleshy that they cannot physically mate, and have to be artificially inseminated. What is shocking is that none of these hapless, deformed animals have to be identified as genetically engineered when sold in the supermarkets or restaurants of the United States. The only way to make sure you aren't supporting this industry is to buy certified organic animal products.

HONORING THE ANCIENT CONTRACT

It is important to understand that it is economics that is threatening the well-being of farm animals—and the health of the humans who eat them. Meanwhile the future of the small family farm hangs in the balance as more and more traditional farmers give up, unable to compete with the soulless, mechanistic, and inhumane practices of the huge multinationals. These giant corporations seek to dominate livestock production on a global scale (just as they try also to dominate seed and crop production, as was discussed in Chapter 4). And so, throughout the world, the old traditional ways are dying out, and the ancient contract between people and the animals who serve them is crumbling.

In view of all this, it is encouraging to know that there are still farmers who care for their livestock, treat them with respect, and thus honor the old contract. Donald Mottram, a British farmer, has such a relationship with Daisy, one of his dairy cows. She always goes to him when he calls to her, leading the rest of the herd. One day, Mottram was savagely attacked by a newly arrived bull. Motrram was knocked to the ground where the bull gored him and stamped on his back and shoulders. Mottram fainted from pain and shock, and when he regained consciousness saw that Daisy, who must have heard his screams, had arrived with the rest of the herd. They were standing around him in a ring and had managed somehow to drive away the furious bull, who repeatedly tried to get at the injured man. The herd maintained its protective circle around the farmer as he dragged himself home. Later he was asked why he thought the cows had protected him.

"Well," he said, "I have treated them reasonably, and they have looked after me in return. People say I am too soft, but I believe you reap what you sow." It is stories like this that help us understand the importance of stopping all the cruelty we inflict on these gentle animals.

A Good Life, a Good Cow

I recently heard this heartening story from the Netherlands, which helps us to remember how the life of a dairy cow can and should be.

"Did we love Trippel because she was so kind, or was she so kind because we hugged her so often?" The Spruit family of Zegveld in the Netherlands does not know. In any case, Trippel was a very special dairy cow. She quite enjoyed attention and she loved to have her photo taken. She produced about 125 kilograms (275 pounds) of milk. But most special of all was that she celebrated her twentieth birthday in March 2005.

A dairy cow reaching the age of twenty years is almost unparalleled. While a nondairy cow's natural life span might be as much as twenty-five years, commercial dairy cows on average live only four to five years. Even at the farm of the family Spruit, the eldest dairy cow is only ten years old. Truus and Theo Spruit care for their dairy herd with love and attention, treating their animals with respect, as if they are humans. They have developed a set of measures to prevent something bad from happening to their beloved dairy cows. It all starts with breeding: Their breeding purpose is a healthy cow that steadily produces milk over a long period of time, instead of the highest amount possible.

"A cow that has to produce milk at such a high level—she is like a top athlete—in general such a cow will be exhausted at an early age," according to Truus. Also, the bulls are selected in such a way that the resulting calves won't be too heavy. That is less risky during calving. The farmer's family also offers a nesting place for swallows. Why? The swallows catch flies, which might transfer mastitis (infections in the udders). And sand is spread over the slippery floor, in order to prevent the cows from slipping and getting hurt. From spring to autumn the cows graze outside, and they get only a limited amount of concentrate feed. (After all, cows are naturally grazers.) Trippel ate virtually everything vegetarian, with apples and strawberries being her favorites. Because of the herd's healthy diet and lifestyle, all of the family's cows produce milk of outstanding quality. Also, their manure doesn't have the bad odor we have come to know in highly intensive dairy systems.

At the age of fifteen, Trippel delivered her last calf. Two years later Trippel was put dry. But grazing in the herd, amongst her fellow dairy cows, she got too fat. So she got a special small paddock to enjoy her retirement, for her final years.

Chapter 6 Farming Animals: Hazards to Our Health

"An industrial meat factory cannot produce a pound of bacon or a pork chop cheaper than a family farmer without breaking the law."

—Robert F. Kennedy, Jr., *Crimes Against Nature*

The modern industrial "farm" has no regard for the wisdom of the true farmer, one who honors his stewardship of the earth, who cares for his animals. Each factory farm "grows" only one kind of animal and these are confined in as small a space as possible and forced to grow and put on weight as quickly and cheaply as possible—in order to make as large a profit as possible in as short a time as possible. In fact, these animal factories are known by the industry as "animal feeding operations."

THEY DON'T EVEN CALL THEM FARMS

In these "operations" animals are fed not with their natural diet but with high-calorie grain that's usually mixed with heavy amounts of corn and maybe a little soy protein. Additionally, it has been common practice to add the ground-up remains of dead animals to cattle feed to increase the protein component. Quite apart from the disease aspect, it seems grotesque to feed cows, who are herbivores, any animal products, let alone forcing them to cannibalize their own species.

The two crops—corn and soy—also happen to be the most commonly grown crops in industrial farms, meaning they are usually grown with heavy doses of chemical fertilizers, pesticides, and herbicides. They are also the most common genetically engineered crops in North America. The emphasis on fast growth thus forces many animals to eat an unnatural diet (the only grains a cow encounters in a pasture are occasional grass seeds) that is typically laced with chemicals, antibiotics, and hormones. Thus every time we eat meat or meat products from these factory farms we are supporting the chemical-dependent agriculture that is poisoning land, air, and water. In addition, we are jeopardizing our health and that of our children—even those people who feel no sympathy for the animal prisoners may be concerned to know that in addition to the hormones and antibiotics in the animals' flesh, all the pesticides, herbicides, and fertilizers used to grow the animals' food also wind up in the animal products we consume. In fact, the pesticide residues are even more concentrated in animal products than in plant food.

DISEASE IN THE SLAUGHTERHOUSE

Almost everyone has heard about *E. coli* O157:H7, the deadly bacteria spread by cattle waste. At least 200 people are reported to have *E. coli* contamination every day in the United States, but health officials believe the number is even higher. Most of the cases can be traced back to sloppy cattle slaughtering and meat processing—allowing fecal matter to be in contact with meat.

Although the United States Department of Agriculture

is supposed to monitor the safety of meat processing, the agency barely penalizes companies for reckless infractions. In 2003, USDA inspectors repeatedly found sides of beef mottled with cow feces while monitoring the Shapiro Packing plant in Augusta, Georgia. The inspectors also discovered a shipment of meat waiting to head out to public schools that was tainted with *E. coli*. (Government-subsidized lunch programs usually supply public schools with the cheapest, worst-quality beef.) Yet the USDA simply issued a warning and allowed the plant to keep shipping out meat, based on promises that they would clean up their plant's operation.

The Food and Drug Administration estimates that 5,000 deaths and 76 million cases of food-borne illnesses occur annually. (Since we don't know the effects of pesticides, antibiotics, and other additives to our meat, this number could possibly be higher.) Yet government policies are so lax around slaughterhouse procedures, many owners of grocery stores and restaurants have decided to exceed USDA standards in order to protect themselves from lawsuits. As a result of an *E. coli* outbreak that killed four Jack in the Box customers in Washington state, the burger chain decided to buy meat only from processors that routinely test for *E. coli* (something not required by the government). It even sends out representatives to inspect the safety of slaughterhouses that sell beef to the plants that grind Jack in the Box's meat. Costco and McDonald's are adopting the same meat-inspecting policies.

BACTERIA

Because animals are packed into such tight, often unsanitary quarters, any outbreak of communicable disease tends to spread very rapidly. Although the USDA insists on random testing for bacterial diseases in processing plants, they cannot catch every infected animal product. It has been estimated that some 80 percent of factory farm chickens are contaminated with *Campylobacter* bacteria, as were almost 95 percent of the pigs slaughtered in the U.K. during 2001. Nearly one in four pigs, chickens, and turkeys are infected with *Salmonella* bacteria. Is it any wonder these two bacteria cause the most common forms of food poisoning? Thousands of people around the world suffer hours of abdominal pain, diarrhea, and fever. And those who are the most vulnerable—children and the elderly or those with weakened immune systems—may die.

Numerous farm animals are also infected with *Yersinia* bacterium. If this gets into a human's bloodstream, it can cause skin rashes, joint pain, and cramping that is so painful it is often confused with appendicitis. And all of these nasty bacteria find the perfect environment for breeding and spreading in the crowded factory farms.

VIRUSES

Then there are the viruses. Mad Cow Disease, or BSE (Bovine Spongiform Encephalopathy), the fatal dementia that began to spread through Britain's cattle in 1985, came from feeding cattle animal products including ground-up meat from "downed" animals (those that fall, often with broken bones, and cannot be used for human consumption).

When people eat meat that comes from cows contaminated with BSE they can become fatally ill with a human form of the dementia called Creutzfeldt-Jakob disease. The FDA now bans animal feeding operations from feeding their cattle dead cow meat, cow's blood, and chicken waste, but it is a hard mandate to monitor. Despite international efforts to contain Mad Cow Disease, it has recently shown up in Canada, the United States, and China.

The new avian influenza (the so-called bird flu) has

spread rapidly through the overcrowded factory farms in Southeast Asia. The first human outbreaks were reported in Hong Kong, but have since been reported in nine other Asian countries. As of May 2005, fifty-three people have been infected, twenty-one of whom have died, and there is a major concern that it is only a matter of time before viruses that have managed to cross the species barrier will mutate into a new potentially lethal form of the bird flu that will wreak havoc on humankind. This, many scientists believe, could supersede HIV-AIDS as the greatest health threat—a global pandemic in the making.

HORMONES

The Bovine Growth Hormones designed to fatten cows quickly also cause painful udder infections. Monsanto, the multinational chemical corporation, makes a widely used Bovine Growth Hormone called Posilac, which comes with the warning that it may cause a number of side effects, including swelling and infected udders. These infections transmit pus, or dead bacteria, and white blood cells into milk, causing a disagreeable taste and unpleasant color. Factory dairies sometimes mix the milk from infected udders with normal milk, so the infected milk with its offensive flavor and color is diluted. U.S. regulations allow milk to contain more pus cell concentrations than any other country in the world—almost twice the international standard of allowable pus.

Routinely pumping growth hormones into farm animals is also linked to the buildup of estrogen in humans. Some scientists believe this explains many emerging biological curiosities, such as why girls are suddenly maturing earlier and why the sperm counts in men are decreasing. These hormones also travel into the waterways through animal waste and are linked to fish developing abnormal sex characteristics, much as the herbicide atrazine is linked to causing bizarre sexual deformities in frogs.

A study in Canada showed that cows treated with rBGH (genetically engineered Bovine Growth Hormone) were 20 percent more likely to be removed from a herd due to health reasons—probably after initially being treated with extra doses of antibiotics (in addition to their antibiotic-spiked feed).

Most American consumers probably do not know that rBGH is banned in the European Union, Australia, New Zealand, and Canada and approval for its use was recently refused by the Codex Alimentarius Commission, the United Nations food safety agency representing 101 nations. It seems strange that America permits the use of rBGH in producing its milk when so many other industrialized nations do not. Today, Monsanto would lose billions if the hormone were banned and is suing organic milk-makers for labeling their milk "rBGH-free," which Monsanto, along with the FDA, claim to be "misleading."

ANTIBIOTIC RESISTANCE

A very serious concern is the routine introduction of antibiotics into animal feed. There are two reasons why antibiotics are routinely pumped into farm animals. One is to protect the frequently anemic animals from diseases due to an unhealthy diet and living in overcrowded, stressful conditions. And the other is because a small dosage of antibiotics seems to help animals grow faster. Every year millions of pounds of antibiotics are fed to livestock, almost eight times the amount given to humans to treat disease. As a result of this routine dosing of animals, many bacteria have built up resistance to the antibiotics on which modern medicine relies so heavily. The antibiotics fed to animals as a prophylactic have now gotten into the human food chain, so that bacterial resistance in humans is increasing at an even faster rate to more and more antibiotics, such as tetracycline, erythromycin, and ciprofloxacin (which was used as an antidote for Anthrax), once thought capable of curing all bacteria-induced diseases.

Poultry workers are already feeling the effects of antibiotic resistance. Chicken handling is classified as one of the most dangerous jobs in the U.S. because of the toxic fumes from waste as well as workers being wounded by terrified birds. But antibiotic resistance has now added a whole new level of risk for the typically low-pay, and sometimes migrant, workers—people who often do not know their rights and never dream of health insurance. Donald Ross used to work at a Virginia chicken factory—weighing them, killing them with a handheld knife in the slaughter room, and hanging them on hooks. One spring day in 2004, Ross accidentally cut the middle finger of his left hand, a wound that might normally heal quickly; instead it swelled into a wound the size of a golf ball. The doctors treating Ross believe his infection was caused by a drug-resistant bacterium in chickens at the plant. Months of antibiotics couldn't heal the infection and eventually doctors had to cut the festering sore right out of his hand. Ross's extreme reaction to the cut and its inability to respond to antibiotics helped launch a public health study, with chicken handlers in the Chesapeake Bay area being tested for antibiotic resistance.

Thus there are now virulent "super-bugs" that can't be treated with any but the most recent and powerful antibiotics. Soon a whole raft of drugs will no longer be useful—to farm animals or to humans. Already a number of people, not at all connected with factory farms, have died when little scratches, such as that sustained by Donald, led to new forms of bacterial infections that spread through the body and failed to respond to any of the antibiotics tried by their frantic doctors. Scientists are working to keep ahead of such resistant strains of bacteria. When a bacterium develops resistance even to these, a nightmare scenario awaits us—graphically described in *The Killers Within: The Deadly Rise of Drug-Resistant Bacteria* by Michael Shnayerson and Mark J. Plotkin. Even though the European Union has finally banned the routine feeding of antibiotics to livestock, the U.S. government continues to fully support this high-profit policy favoring American meat and pharmaceutical companies.

ANIMAL WASTE: THE CYCLE OF POLLUTION

Excessive, concentrated animal waste harms our environment in many ways: It contributes to the greenhouse gases that cause global warming, adds to the problem of acid rain, contaminates our waterways and oceans—and it creates horrendous "smell pollution."

As we said, in small-scale farms, where cows, pigs, and hens graze and roam the fields, their waste provides natural fertilizer for the farm's ecosystem. But a factory-style animal feeding operation, with hundreds and even thousands of animals crammed together in a small space, inevitably creates much more manure than the factory's land can naturally absorb. The amount of waste produced by farm animals in the U.S. is estimated to be more than 130 times greater than human waste. But unlike human waste, factory farm waste doesn't go through a sewage treatment plant.

Everyone who has ever changed a litter box, or cleaned out a wet diaper bin, is only too familiar with the potent presence of ammonia in animal waste. Imagine how intense the ammonia is where numerous animals are raised in confinement. Indeed, unless there is good ventilation, the volatile ammonia accumulating in an indoor poultry factory can actually damage human—and chicken—eyes. Factory-style poultry farms located near the Chesapeake Bay pour millions of tons of ammonia-soaked waste into the fragile waterway every year. The high-nitrogen ammonia acts as a nutrient for algae, so that at certain times of year they proliferate to the extent that they create "dead zones" where no fish or plant life can survive. The last time it was measured, the Chesapeake's dead zone took up 40 percent of its bottom waters—the largest dead zone in the nation. And of course this situation is not unique to the Chesapeake—dead zones and fish poisoning occur wherever there is intensive animal farming. The Gulf of Mexico has a vast dead zone partly caused by animal waste spillovers from animal feeding operations. In 2002 it was 8,500 square miles—about the size of Israel.

PIG BUSINESS IS BIG BUSINESS

Pigs, America's most popular meat source, create ten times more waste per day than humans. Current federal regulations stipulate that manure from factory farms—or Confined Animal Feeding Operations—need only be contained in "ponds" lined with "compacted clay"—a substance that expands and contracts with changes in moisture, and in the process can crack. Then the liquid manure leaches directly into the groundwater, wells, and waterways. Although some states have stricter policies, many regulating agencies have a difficult time monitoring what is going on in the ponds—indeed, some agencies have a hard time keeping track of the *number* of ponds any given feedlot has on-site, not to mention their quality.

This stinking waste is a major health hazard for humans and animals alike. In his book *Crimes Against Nature,* Robert F. Kennedy, Jr., describes it as "a witch's brew of nearly 400 toxic poisons, including heavy metals, pesticides, hormones, deadly biocides, and dozens of disease-causing viruses and microbes." Hog waste seems to have fostered outbreaks of a previously unknown microbe, *Pfiesteria piscicida*, in U.S. coastal waters. This microbe, known as the "cell from hell," kills millions upon millions of fish. In one six-week period in 1991, it killed an estimated one *billion* fish in the Neuse River in North Carolina. It also affects humans, causing "pustulating lesions that won't heal, severe respiratory illness, and brain damage in humans who handle fish or swim in the water." Also as much as 25 to 75 percent of the antibiotics fed to animals in factory farms are released with urine and feces, providing great opportunities for bacteria to develop resistance to them. Indeed, one study showed that 2.1 percent of bacteria found naturally in the soil of one field showed resistance to the antibiotic used in a feedlot two kilometers away.

In North Carolina hog waste became a major issue in 1999 when the torrential rains of Hurricane Floyd pounded the region. Besides the pollution caused by the rotting carcasses of the 100,000 drowned pigs, the stinking, liquid manure escaped from the holding lagoons and poured directly into the groundwater and rivers. The brown stain of pollution extended for miles, contaminating drinking water all along the coastal South and finally dumping its poison into the Atlantic Ocean.

Even under normal weather conditions, manure from industrial hog farming is the number one pollution threat to the rivers and waterways of the U.S. And the stench from these farms is overpowering, and so noxious that it causes severe headaches and nausea so that the people in neighboring households are forced to stay indoors most of the year. Workers have actually died from falling into a manure lagoon because the gases are so lethal.

"The hog barons' business model relies on the assumption that they can evade prosecution for these crimes by improperly influencing government enforcement officials," writes Kennedy. He believes that corporate meat producers "intentionally locate their meat factories in poor and minority communities where they can crush and muzzle opponents, harassing neighboring farmers for complaining about odors or water pollution or for participating in public hearings." Unfortunately, these corporations have the money and the political capital to ignore (or reformulate) the standards that have been developed to protect our precious planet. They even infiltrate state legislatures and in some states, including Iowa, North Carolina, and Michigan, have stripped local officials of their decision-making powers so that hog factories cannot be zoned out by health officials.

And these people, even if they protest, cannot afford to pay lawyers to fight for social justice. Which is why Robert Kennedy decided to challenge the Smithfield hog facilities in North Carolina. Eventually he was able to win his case against four of the Smithfield facilities that were dumping toxic waste from their pork factories onto the soil without a permit. The permit requires that the waste first be treated to ensure it does not pose a risk to the environment or public health. When Kennedy brought a case against them, for ignoring the requirement and violating the Clean Water Act, Smithfield argued, unsuccessfully, that it should be given the same exemptions that are given to small individual farms. This seemed like a major victory for Kennedy. However, a powerful group of pork industry leaders, lawyers, and lobbyists met to strategize their way around the law. They drafted a new set of regulations and, through brilliant political maneuvering, persuaded the EPA to adopt new regulations similar to those prepared by Smithfield's team. And these new regulations removed stringent environmental safeguards that had been part of the Clean Water Act. No longer can Smithfield, or any other meat industry company, be held accountable for the waste its farms produce. No longer is it necessary for the meat industry to monitor groundwater contamination levels. Kennedy points out that the Clean Water Act suffered because of a political environment that permits corporate irresponsibility. Sadly, it seems that the political will to address human health issues, as well as to enforce environmentally sustainable agriculture, is weak.

Bushmeat: The Hazards of Eating Wild Animals

Hunting "for the pot" has always been part of rural life worldwide. In Africa, people living in the rain forests have hunted for meat from the dawn of human history. For hundreds and hundreds of years they lived in harmony with their forest world, practicing subsistence hunting—killing just enough to feed their families and villages.

Today, however, the commercial bushmeat trade worldwide has become so large that animals are literally being eaten to the brink of extinction. Indeed, Miss Waldron's Red Colobus, a beautiful black and red monkey that once lived in troops in the rain forest canopies of West Africa, has officially been declared an extinct species. If the bushmeat trade in endangered species isn't stopped soon, the great apes—the gorillas, bonobos, and my beloved chimpanzees—may be all but gone from the great forests of the Congo Basin within the next ten years or so.

Most people are shocked to learn that tons of bushmeat (often antelopes, but also monkeys) are shipped out of Africa to Europe and America, mostly to cater to the cultural preference of African communities living abroad. In one year, customs agents impounded more than fourteen tons of illegal bushmeat at Heathrow Airport alone. Along with monkey and antelopes, they found meat from anteaters, bats, and the little smoked legs of tortoises.

It is not only Africa's wild animals that are threatened with extinction: Mammals, reptiles, and amphibians of many kinds are hunted and sold for food throughout South America and Asia. And birds are slaughtered in their thousands. In Central and South America curassows, chachalacas, and guans (all large, turkey-sized, and very palatable birds) are the first to go as hunters penetrate deeper and deeper into the forests. Songbirds are also killed for food in many other countries such as India, China, Italy, Spain, France, Greece, and Cyprus. In 2003 French conservationists tried to persuade the European Court of Justice to ban the shooting of wild birds during nesting, rearing, and breeding. Shockingly their plea failed—although the Court did rule that hunting was to be supervised, and only small numbers of certain birds could be shot. Countless fish species, hunted to the brink of extinction, can be seen as the bushmeat of the water.

SARS, HIV, EBOLA: HUMAN HEALTH AND BUSHMEAT

The bushmeat trade is not only threatening many animal species with extinction but in some places is directly affecting human health. There is now sound scientific evidence indicating that the HIV-AIDS pandemic originated when humans were infected by a retrovirus carried by chimpanzees, the chimpanzee simian immunodeficiency retrovirus, which does not lead to symptoms of illness and is perfectly benign so long as it remains within the host species. However, at some point it jumped the so-called "species barrier" (differences in blood, immune system, and so forth that prevent dogs getting sick with polio or humans with distemper) into a human being, where it mutated into the HIV retrovirus that causes AIDS. This happened in two different parts of Africa, giving rise to HIV-1 and HIV-2. How did this virus jump from chimpanzees to humans? It has been suggested that it may have been the result of bushmeat hunting, as people became contaminated with the blood of infected individuals during butchering for sale—the "cut hunter" theory.

More recently, the world was alarmed at the prospect of a pandemic as a result of the SARS virus. The origin of this was found to be the palm civets (mongoose-like creatures) that are much prized as meat in China. Many of the vendors keeping and selling civet meat in markets where conditions are both unsanitary and inhumane were found to have built up resistance to the virus, presumably as a result of previous infection with SARS.

In 2005 the newspapers reported that a new Ebola-like disease was discovered in Angola that claimed the lives of 112 people. Three of these deaths and two of those reported sick were in Lusaka; they had come from a province along the Congo border where the outbreak was discovered. And many were speculating there was a link to the flourishing trade in bushmeat.

The potential human health hazards posed by the increasing trade in wild animals for human food are daunting. One of the best ways to discourage the killing and eating of bushmeat is to evoke empathy for the animals themselves. There are efforts to teach children to better understand and value their wildlife. In the countries of the Congo Basin some nongovernmental organizations (NGOs) are offering this curriculum to schools. Many African cultures greatly respect the apes and frown on their consumption, so local villagers, hunters, and elders are encouraged to integrate their own legends about apes (as well as other animals) into the curriculum. The Jane Goodall Institute's (JGI) Roots & Shoots program for young people is working with other groups, such as

Great Ape Survival Project (GRASP), the Ape Alliance, and the Diane Fossey Gorilla Fund, to try to create a new generation of citizens who love and want to protect the great apes and other animals.

In Congo-Brazzaville, in the very heart of the bushmeat trade, JGI is caring for more than a hundred orphan chimpanzees, most of whose mothers have been killed for food. We have been accused of "wasting" money that could have been better spent on protecting wild chimpanzees and their vanishing habitat. But so many of the local people, especially the children, who visit the sanctuary go away saying they will never eat chimpanzee meat again. Some say they will never again go to a restaurant or house where it is served. Thus our orphans are truly ambassadors, helping the overall effort to prevent the extinction of their species.

Chapter 7 Helping Farm Animals Live Better Lives

"The greatness of a nation and its moral progress can be judged by the way its animals are treated. . . . I hold that, the more helpless a creature, the more entitled it is to protection by man from the cruelty of man."

—Mahatma Gandhi

Is there anything we can do about the cruelty and health hazards described in the previous chapters? Of course there is. First we have to let more people know what is going on. Even now, despite the revelations of animal rights advocates, most people still have absolutely no idea about the suffering that goes on, day and night, out of sight. And unfortunately, only too often, they prefer not to be educated. "I'm too sensitive," they tell me. "I love animals. I can't bear to think of them suffering. Please don't tell me about it." But only when we know the full horrific details will we do something to improve the way of life for the animals we care about. We must not stand by passively and watch compassion in farming disappear, bury our heads in the sand, and carry on with business as usual.

The more people become aware of animal pain, the more they feel compelled to do something to help. Fortunately, there are some influential voices raised on behalf of our farm animals. Senator Robert Byrd deserves full credit for his strong stance on improving conditions in the animal-killing factories of the United States. He delivered a landmark speech to the U.S. Senate in July 2001, in which he railed against the barbaric institutionalized cruelty that is inflicted on food animals in the slaughterhouses, and called for the USDA to introduce measures that would end this cruelty. He pointed out that animals suffer pain, and made a plea for "respect for all life . . . and for humane treatment of all creatures." Senator Byrd is a true humanitarian who is not afraid to make his views known. I salute him.

Governor Arnold Schwarzenegger helped California become the first state to do something about the terrible plight of geese and ducks raised for foie gras. In 2004 he signed a law that makes it illegal to forcefully enlarge bird livers "beyond normal size." After the law passed, the animal rescue organization Farm Sanctuary held a celebration party, drawing out famous animal lovers, such as James Cromwell (the weathered-faced farmer in the movie *Babe*), who helped serve crackers and "faux gras"—a popular vegan pate made from tofu, seiten, and lentils.

Another celebrity, Mary Tyler Moore, is one of the leading animal rights activists in the U.S., insisting that farm animals stop being treated as "tools of production." She is the honorary chair of the Farm Sanctuary's Sentient Beings Campaign, speaking out all over the world on behalf of the many animals who cannot speak for themselves.

But it is not only influential people whose voices are needed—all our voices are needed. Each one of us can make a difference. After one million people signed a petition and brought it to the European Union, the heads of government finally agreed to recognize farm animals as sentient beings, which means that inhumane farming methods, such as foie gras production, and the raising of calves in veal crates, will soon be banned in over a dozen countries in Europe. The United Nations is now paying attention to this swelling movement and recently issued a few reports, recognizing the sentient rights of farm animals. Every day, slowly but surely, we, the people, are helping to change the world.

USE YOUR CONSUMER POWER

While it's always a good idea to keep putting pressure on the government to hold the agriculture industry and factory farms more accountable for the harm they cause, it is probably more effective to focus on those who are actually marketing animal products. Imagine the global changes that would take place if everyone who cares made their concerns known by telephoning, writing, or simply talking with store managers of the meat-purchasing titans, such as Wal-Mart, Wendy's, or Kentucky Fried Chicken. No matter how big a business, it cannot remain successful unless customers continue to buy its products.

Thanks to consumer demand, there is a rapidly growing number of dairy, egg, poultry, pig, and cattle farmers, as well as fishermen, who are dedicated to the physical and psychological health of their animals, and the sustainability of the planet, as well as the gratification of their customers. We can now go to just about any grocery store and find products from animals who were raised humanely, safely, and in ways that caused little or no harm to the environment.

If you have any doubt about consumer power to move mountains, keep in mind that it was customer pressure that helped convince McDonald's to request that its suppliers phase out the use of antibiotic growth promoters. While this policy doesn't stop antibiotic dependence in the factory farms (the industry still needs antibiotics to treat diseases that stem from poor diet and unsanitary, over-crowded conditions), it is a step in the right direction. And in 2005 Oregon's Tillamook Creamery Association forced all of its dairies to abandon the use of Monsanto's Posilac after receiving a barrage of customer queries and complaints about the effects of Bovine Growth

Hormones. If more and more consumers make their concerns known, we shall soon see significant changes.

HOLD RESTAURANTS ACCOUNTABLE

We can share our concerns with restaurants. When my mother Vanne heard about the torture of calves for veal production, she instantly stopped ordering Vienna schnitzel—even though it had always been her favorite item from the menu. (This was before she became a vegetarian.) It was public pressure from people like her, the loss of customers and potential customers, as well as the presence of protesters lobbying outside the restaurants, that made a big difference in the lives of thousands of veal calves. Even people who continued to eat veal learned to ask for "pink" veal—in other words, the meat of calves who had not been kept in tiny crates and deprived of iron.

In the U.S., customer pressure has helped more than 400 restaurants sign pledges not to serve pâté de foie gras. And in the Netherlands, there is a major campaign to raise awareness about the torture implicit in making this delicacy. Restaurants that agree to stop serving it have signs pasted over the door by the activists, making it easy for those who care about the well-being of animals to decide where to dine.

TALK TO YOUR GROCER

When I told my mother Vanne about battery farms for hens she was in her late seventies. She was horrified! And so the next time she went into one of the big supermarkets in our hometown, Bournemouth, she looked for a sign indicating free-range eggs. She couldn't find one, so she went over to a young assistant and asked if they had any.

"What's a free-range egg?" asked the girl. My mother explained that it was one laid by a hen who could move about and scratch around on the ground. "Don't all hens do that?" asked the assistant. So my mother explained about battery farms, the tiny cages, the de-beaking and how the blood oozed from the mutilated beaks, and so on. The young woman looked horrified, and a small crowd of customers gathered around to listen.

The assistant manager was called. He led the trouble-making customer into his office—where Vanne went over the facts again. The following week, and in all subsequent weeks, free-range eggs were on sale in that store. So my mother did the same in other stores. It just shows what one determined person can do. And when this effort is multiplied, we begin to see a real difference.

Already there is a difference. It is now easy to buy not only free-range eggs, but organically produced free-range eggs, in the supermarkets of the U.K. and most of continental Europe. And it is becoming easier to buy them in North America. It is the same with organic chicken, beef, and pork. And it is the same with milk. Every time we buy an organic dairy product we are taking a step toward improving the lives of dairy cows. For one of the requirements for certification of organic milk is that the cows must have access to pasture for at least part of the year. Every pint of organic free-range milk we buy makes a difference, helping to rescue cows from a life of tortured confinement and the hideous side effects of Bovine Growth Hormones. The same is true when we buy organic free-range butter, cheese, and yogurt. (And, incidentally, when we buy animal products from people who have treated animals humanely, we not only help good-intentioned farmers, we also help the environment and our own health.)

There's a legendary story about John Mackey, the founder and CEO of Whole Foods, the largest organic and natural foods grocer in the world. One day he was approached by Lauren Ornelas, a representative from Viva!USA, an animal rights group, who complained about the inhumane treatment of the ducks that were sold "ready-to-cook" at Whole Foods. At first Mackey was annoyed by her complaining. His grocery chain was already trying to eliminate chemical toxins and hormones in its meat and seafood, what more did she want? But Ornelas was both persuasive and persistent, and finally Mackey decided to look into her concerns.

He spent three months educating himself about the treatment of factory farm animals. His solo tutorial changed the way Whole Foods purchased meat. Nowadays the grocery chain will only buy animal products from suppliers that have treated their animals with dignity and allowed them sentient rights before slaughter. This meant that his California duck supplier of the past nine years had to make some changes in order to keep Whole Foods as its retailer. The supplier stopped trimming the ducks' bills and even built a special pool so the ducks had the opportunity to swim during their lifetimes. The tutorial also changed Mackey's life. He became a strict vegan, no longer eating any food from animals, including dairy products.

I met John Mackey early in 2005. We had dinner at a vegetarian restaurant in London and discussed the whole sorry situation of the way farming is conducted by the multinational corporations. Certainly John's store, Whole Foods, is making a difference, and the more people shop at his store, and the more successful he becomes, the sooner other stores will follow his example.

READ LABELS CAREFULLY

When you buy "certified organic" animal products you have a reasonable guarantee that the meat, poultry, eggs, or dairy products you are about to eat came from animals that weren't fed hormones, antibiotics, and animal by-products. The animals also have to be raised on food that doesn't contain synthetic pesticides and fertilizers, or food that was genetically engineered or zapped with ionizing radiation. In most cases, the "certified organic" label also guarantees that the animal was "free-range," meaning that it was allowed access to fresh air, exercise, and pasture.

But labels can be deceptive, and one needs to study them carefully. For example, "free-range" alone, without the organic label, doesn't necessarily mean it's "organic" as well. "No hormones added" could be deceptive because it doesn't automatically imply that no antibiotics, pesticides, or other chemicals were used—such as the chemical dye sometimes given to factory farm hens to give their egg yolks the "natural" yellow look that is induced by sunlight in free-range hens. And beware of labels such as "all natural," which have no clear meaning.

BUY GRASS-FED ANIMAL PRODUCTS

One promising trend in sustainable animal farming is actually a return to the age-old solar-powered system of letting animals get their nourishment from the pasture. Because grass-fed livestock get natural food, fresh air, and sunlight, they are normally far healthier than factory-farmed animals. Without the stress of the crowded conditions they are more resistant to disease and thus don't need all the antibiotics and other boosters that are given to factory-farmed animals.

Grazing cattle, provided the herd is not too large for the space allotted, can actually maintain grassland by preventing trees and shrubs from taking over. As grassland takes over from cultivated fields, especially fields of a monoculture, soil erosion can be reduced by as much as 93 percent. And as their hooves step through the grasses, they help to disperse their seeds. At the same time the cows fertilize the soil with their manure. A growing number of farmers are practicing "green grazing" or "conservation grazing," a type of land management that is specifically designed to restore parts of the prairie from corn and soybean fields to grazing land. A few farmers are even ranching buffalo in some parts of the prairies.

Of course, if the population of the U.S. continues to grow, and if people continue to demand a high-meat diet, there would not be enough grassland to sustain America's 100 million cattle, let alone all the other farm animals. And it would be horribly unethical to place any further demands on other countries—already vast areas of the Brazilian rain forest have been destroyed in order to create grazing for cattle. Therefore, for environmental as well as health reasons the most important decision for the meat-eater is to eat only very small amounts of meat. It is desperately important to cut back on the current greedy and unsustainable overconsumption of meat that is the norm for many around the world.

Health Advantages of Grass-Fed Animals

Few people are aware of the health benefits of eating products from animals that were allowed to graze. For instance, meat or dairy products from grass-fed animals are lower in saturated fats and harmful cholesterol, and they are higher in vitamin E and the beneficial omega-3 fatty acids—which most people get from flaxseed oil, fish oil supplements, or eating wild salmon. A steak that comes from a grass-fed cow can have as much as six times the omega-3 oils of a feedlot steak, according to research from the University of Hawaii. Meat and dairy products from pasture-raised cows also contain six times more CLA (conjugated linoleic acid), and studies (at the University of Bristol in England, Cornell University, Penn State, and the University of Utah) have shown that CLA slows the progress of some types of cancer and heart disease.

Granted, the tenderness of grass-fed beef isn't always dependable. But unlike the uniform taste of corn-fed animals, meat and dairy products from grass-fed animals tend to take on slightly different flavors, almost as if they represent the different varieties of grass and herbs the animals ate as well as other subtle nuances of living a more natural, free-range life.

There are a number of ranchers in the Central Plains area who have begun ranching bison—generally called buffalo. And it is probably true that buffalo meat is one of the healthiest for us. Grass-fed buffalo meat is also high in essential fatty acids with a good ratio of omega-3 to omega-6. It contains high levels of CLA (which can also reduce body fat and increase muscle mass). Buffalo meat is 98 percent fat-free and has 35 percent more protein than beef. Buffalo contains high levels of vitamins and minerals such as iron and twice as much beta-carotene as grain-fed meats. It also has less fat and cholesterol than chicken, turkey, or halibut. Amazingly, most meat-eaters can reduce their LDL cholesterol by 40 to 45 percent over a six-month period by eating five ounces of grass-fed buffalo meat four to five times a week. The buffalo is one of the only land mammals that do not contract cancer. They actually carry a cancer-preventive enzyme.

SUPPORT THE RIGHT FARMERS

George Vojkovich farms the Skagit River Ranch, situated in a rich agricultural valley in the far northwest corner of Washington state. For most of his life he did what most farmers do today. He bought chemical fertilizers and pesticides and other farm products from an "agronomy center" that was, of course, owned by a chemical corporation. Staff from the agronomy center always had "just the right chemicals" to improve the quality of soil and eliminate insects, pests, and so on.

One day, the forty-four-year-old farmer suddenly started having irregular heart rhythms. After being hospitalized and put on blood thinners, George thought his days were numbered. But his doctor told him that he saw this kind of heart problem all the time, and he believed it was caused by exposure to toxic chemicals. "Get rid of the chemicals and you'll get rid of the problems."

The first thing George did was switch to an organic diet. His heart problems also convinced him and his wife, Eiko, to stop putting chemicals on their soil. They were determined to create a farm with the healthiest, most nutrient-rich soil possible. It took seven years of intensive composting to get the soil and grassland organically fertile enough to sustain farm animals and obtain USDA organic certification.

Like the farms of yore, Skagit River Ranch is populated with cows, pigs, and chickens—all of whom get plenty of friendly contact with George's farmhands and family, including his nine-year-old daughter, who is especially fond of the pigs. Nowadays, all the animals are pasture-raised, meaning they are rotated around the property where they feast upon the ranch's organic heirloom grasses. As the animals rotate through fields, George follows behind them, planting rye and clover. But he also plants a smattering of organic carrots and turnips, simply because the pigs love finding these delicious and highly nutritious treats. Compared to their factory-raised cousins, George's animals have longer lives. Because they are allowed to roam and exercise and because they are not given hormones and antibiotics to accelerate their growth, it takes them almost twice as long to reach maturity as a factory-farmed animal.

After several years of farming without chemicals, George's heart fibrillations stopped and he became noticeably healthier. And after several years of developing a more intimate relationship with the soil and animals, other remarkable changes took place. Perhaps the most extreme change in George's farming methods can be seen at the time of slaughter. For one thing, a USDA-certified organic "slaughterhouse"—a mobile trailer—comes to his farm so that none of his animals have to endure the stress of a journey. And, even more remarkable, George now offers his animals a goodbye blessing before they are killed.

"Years ago if you asked me if I said a prayer before I slaughtered an animal, I would have laughed at you," says George. "But now I understand that animals, humans, plants—we are all part of this planet. We are all part of the whole, and it's greater than we can imagine."

So what does he say to his animals as he leads them to slaughter? There's not a set speech, says George, but it usually goes something like this: "Hey, I'm sorry, but this is the way it is. I thank you for helping us here and doing your part and being such a good animal. And I know that where you are going isn't by your choice. It's by our choice, and please forgive me for that."

Many of his customers find it a relief to eat animal products that are raised so organically and ethically. "But some of my customers don't give a darn about all that," admits George. "They just know it tastes great." In a 2004 national taste competition among growers of grass-fed beef, sponsored by Eatwild.com and *Stockman Grass Farmer*, George's beef won first prize for flavor and tenderness.

One of the toughest challenges facing small family animal farmers such as George is getting their products into the hands of appreciative customers, since most of the packaging and distribution of America's meat is handled by a few large impersonal corporations. This is where Niman Ranch, perhaps the most progressive and innovative meat company in the U.S., comes into the picture. It began almost thirty years ago as a small-scale operation, supplying healthy and humanely raised beef to organically minded customers in California's Marin County. Then, as people became more educated about the cruelty and health hazards of factory-style animal farming, the ranch's popularity and mission began to grow. Nowadays, although Niman Ranch still raises cattle on the original property, it has expanded to become the packager and distributor of beef, pork, and lamb that comes from over 300 small-scale farms. Not only will you find Niman Ranch products in chain stores, such as Whole Foods and Trader Joe's, you can also order their products off the Internet.

An increasing number of people want a personal connection with the land, and with the farmers who care for the animals whose flesh they eat. If information about the lives of the factory-farmed animals whose meat is sold in the supermarkets was spelled out, it is doubtful if any would be sold. But the story behind Niman Ranch meat serves as its greatest selling point rather than a dirty secret to hide from guileless consumers. Each package states that the meat came from a farm where the animals are treated humanely, raised on land that is cared for in a sustainable system, given food with no animal by-products or waste, and never given growth hormones. Nor are they given antibiotics for anything other than to treat sickness. (And if the animals are given antibiotics, they are not used for meat later on.)

Niman Ranch only does business with family farmers who own or lease their farmland. The Niman Ranch Web site (www.nimanranch.com) is a masterpiece of customer relations, even offering pictures and written profiles of its pig farming families. Some pictures show parents with their young children, others feature couples, or siblings, or fathers and sons. Many families proudly pose with pale pink piglets in their arms and weathered barns and farmhouses in the background. When saying what they like best about pig farming, farmers often share their love of the farrowing season—when the piglets jostle for teats, bonding with their mothers, then making friends with the rest of the clan. Almost all these farmers grow their own animal feed, and make a point of letting customers know that they would *never* feed their hogs genetically engineered corn or soy.

Niman Ranch Web site visitors can also read how the hogs root and rotate through fields that are often fertilized with compost mixed with hog manure: "Although this traditional system requires more hard work than the modern confinement alternative, it produces little to no odor, helps preserve water supplies, and sustains the land and community for future generations." What a startling contrast to the industrial hog farms that can't even contain their overflowing manure, much less figure out how to sustain their land with it.

Niman Ranch follows humanitarian protocols developed by the Animal Welfare Institute. The farmers not only socialize with their animals in the pasture, but actually accompany them to slaughter. When pigs go to a slaughterhouse, workers comment on how much more relaxed these animals are than those brought in from factory farms. This lack of fear is also one of the many reasons Niman Ranch's meat tastes so good, according to connoisseurs, and is preferred by fine chefs all over the nation. (Adrenaline released by fear at slaughter is said to cause meat to become drier and tougher.)

In light of so many food scares, customers are especially reassured by Niman Ranch's attention to health and safety. When a cow (originally from an Alberta, Canada, dairy farm) was discovered to have Mad Cow Disease in Washington state in 2004, Niman Ranch's business at Whole Foods and Trader Joe's immediately went up 30 percent compared to the previous year.

How Much Does It *Really* Cost?

Before industry took over animal farming, chicken was more of a luxury rather than a cheap fast food. Herbert Hoover's 1928 campaign slogan, "A chicken in every pot," came from his promise that every American would finally be wealthy enough to afford a nightly chicken dinner. Since then, animal feeding operations took over production, cranking out animal products at a high volume and low cost—factory-raised chicken, beef, and pork are all so cheap most Americans can eat them in abundance.

One of the arguments against pasture-raised animal farming is that the small-scale, environmentally sustainable operations can't produce animals as cheaply as the factory farm operations. Someone has to make up for the price difference, and that means the consumer. A whole, pasture-raised broiler chicken who was bred on the cornucopia of insects and worms in rich organic soil might cost about $15. A factory-farm chicken will cost about half the price. Grass-fed beef is also two or even three times more expensive than meat from an animal feed operation.

However, an investigation into the facts suggests that, if all factors are taken into consideration, the true price for a factory-farmed animal is just as high, or even higher. Only this true cost is hidden from us. For instance, the price paid over the counter for a Big Mac does not reflect how much the government has already taken from the taxpayer for the government subsidy of corn crops—the staple of the feedlot diet. The cost of a bucket of Kentucky Fried Chicken never factors in the price paid for the environmental pollution that factory farming causes. And the price on the shrink-wrapped grocery store meat doesn't factor in the hidden cost of treating the illnesses and diseases that are caused by eating the meat of factory-farmed animals. If we think in these terms, the true cost of a pound of free-range organic meat is comparatively less than what we pay for a pound of factory-farmed meat.

Even people on low incomes, if they truly understood the situation, could still afford the extra money it would take to buy free-range organic meat by simply eating less of it.

ONE THREAD AT A TIME

One of the most knowledgeable and outspoken critics of contemporary factory farming is University of California at Berkeley journalism professor Michael Pollan. He likens consumer choices to pulling single threads out of a garment. We pull a thread from the garment when we refuse to purchase eggs or meat from birds who were raised in confinement, whose beaks were clipped so they could never once taste their natural diet of worms and insects. We pull out a thread when we refuse to bring home a hormone-fattened turkey for Thanksgiving dinner. We pull a thread when we refuse to buy meat or dairy products from cows who were never allowed to chew grass, or breathe fresh air, or feel the sun warm their backs.

The more threads we pull, the more difficult it is for the industry to stay intact. You demand eggs and meat without hormones, and the industry will have to figure out how it can raise farm animals without them. Let the animals graze outside and it slows production. Eventually the whole thing will have to unravel.

If the factory farm industry does indeed unravel—and it must—then there is hope that we can, gradually, reverse the environmental damage it has caused. Once the animal feed operations have gone and livestock are once again able to graze, there will be a massive reduction in the agricultural chemicals currently used to grow grain for animals. And eventually, the horrendous contamination caused by animal waste can be cleaned up. None of this will be easy.

The hardest part of returning to a truly healthy environment may be changing the current totally unsustainable heavy-meat-eating culture of increasing numbers of people around the world. But we must try. We must make a start, one by one.

HELP RESCUE ANIMALS

Several years ago, when I was traveling on a train from London to Cambridge, I met a woman who had rescued twenty hens from a battery farm. She got them for a couple of pennies each: They had reached that stage in the factory farm hen's life when they are good for nothing more than chicken soup or fertilizer. Almost entirely without feathers, weak from their lifetime in the cramped hell of their "farm" prison, they huddled miserably on the ground when she set them down in the fenced-in portion of her estate she had prepared for them.

"It was several days before they even tried to walk," said my traveling companion, "but they learned in the end." She told me how much difficulty they had trying to peck around, the way hens do, because they had been debeaked. "Probably hurt them, poor things." After several weeks of space, sunshine, and good grain with household scraps added for variety, they had grown back their feathers. And eventually they rewarded her by laying a few eggs.

I asked her why she had done this: It was after seeing some of the photos published by the Humane Farming Association. "I felt so damned sorry for them." She told me she had rescued batches of wretched, exhausted birds every year for the past three years.

What is so extraordinary is that a few weeks after that meeting I encountered an old couple who lived in a small house with a small garden—what's called a yard in America—who, for the same reason, did exactly the same! They could only take three or four at a time, but they wanted to try to give at least these few a chance to know freedom before they died. If I lived on a farm I would want to rescue hens, as well as pigs from their hellholes, little calves from their veal crates—and so many more. I applaud those who do. Although there are not many of us who can create homemade havens, there are more and more sanctuaries springing up around the world for animals in need of care, whether for injured wildlife, rescued farm animals, or ex–circus, zoo, or medical laboratory animals. Farm Sanctuary, a national rescue and adoption network, was started by two caring activists in 1986 when they first rescued Hilda, a sheep who had been abandoned and left for dead on a stockyard "deadpile." Today, Farm Sanctuary is the largest farm animal rescue and protection organization in the U.S. One of their many programs is the Adopt-a-Turkey Project, which encourages people to adopt or sponsor a Thanksgiving turkey rather than eat one. All of the sponsored turkeys are treated to a Thanksgiving feast of stuffed squash, cranberries and pumpkin pie at one of the sanctuary's shelters.

Because people are becoming increasingly aware of animal pain, they are feeling more and more compelled to do something to help. We have listed some sanctuaries and animal rights organizations at the back of the book and hope that you would support them with your dollars or volunteer efforts.

Most people are basically decent; most people do not like to think of animals suffering at our hands; most people want to do their bit to make this a better world for all—only they don't always know quite what they can do. So let us join forces, let us not turn away from this torture of millions of animals. Each one of us can do our bit. We can change the way we eat. We can refuse to buy animal food produced by inhumane farming methods, thus lobbying for change with our purse. We can contribute to the lasting welfare of animals living in sanctuaries. And we can help to spread awareness of what is going on.

Excerpt from speech delivered by Senator Robert Byrd, July 9, 2001, United States Senate

"Our inhumane treatment of livestock is becoming widespread and more and more barbaric. Six-hundred-pound hogs—they were pigs at one time—raised in two-foot-wide metal cages called gestation crates, in which the poor beasts are unable to turn around or lie down in natural positions, and this way they live for months at a time.

"On profit-driven factory farms, veal calves are confined to dark wooden crates so small that they are prevented from lying down or scratching themselves. These creatures feel; they know pain. They suffer pain just as we humans suffer pain. Egg-laying hens are confined to battery cages. Unable to spread their wings, they are reduced to nothing more than an egg-laying machine.

"Last April, the *Washington Post* detailed the inhumane treatment of livestock in our nation's slaughterhouses. A twenty-three-year-old federal law requires that cattle and hogs to be slaughtered must first be stunned, thereby rendered insensitive to pain, but mounting evidence indicates that this is not always being done, that these animals are sometimes cut, skinned, and scalded while still able to feel pain. A Texas beef company, with twenty-two citations for cruelty to animals, was found chopping the hooves off live cattle. In another Texas plant with about two dozen violations, federal officials found nine live cattle dangling from an overhead chain. Secret videos from an Iowa pork plant show hogs squealing and kicking as they are being lowered into the boiling water that will soften their hides, soften the bristles on the hogs and make them easier to skin.

"I used to kill hogs. I used to help lower them into the barrels of scalding water, so that the bristles could be removed easily. But those hogs were dead when we lowered them into the barrels. The law clearly requires that these poor creatures be stunned and rendered insensitive to pain before this process begins. Federal law is being ignored. Animal cruelty abounds. It is sickening. It is infuriating. Barbaric treatment of helpless, defenseless creatures must not be tolerated even if these animals are being raised for food—and even more so, more so. Such insensitivity is insidious and can spread and is dangerous. Life must be respected and dealt with humanely in a civilized society.

"So for this reason I have added language in the supplemental appropriations bill that directs the Secretary of Agriculture to report on cases of inhumane animal treatment in regard to livestock production, and to document the response of USDA regulatory agencies. The U.S. Department of Agriculture agencies have the authority and the capability to take action to reduce the disgusting cruelty about which I have spoken. Oh, these are animals, yes. But they, too, feel pain. These agencies can do a better job, and with this provision they will know that the U.S. Congress expects them to do better in their inspections, to do better in their enforcement of the law, and in their research for new, humane technologies. Additionally, those who perpetuate such barbaric practices will be put on notice that they are being watched.

"I realize that this provision will not stop all the animal life in the United States from being mistreated. It will not even stop all beef, cattle, hogs, and other livestock from being tortured. But it can serve as an important step toward alleviating cruelty and unnecessary suffering by these creatures. . . .

"Thus, Mr. President, God gave man dominion over

the earth. We are only the stewards of this planet. We are only the stewards of His planet. Let us not fail in our divine mission. Let us strive to be good stewards and not defile God's creatures or ourselves by tolerating unnecessary, abhorrent, and repulsive cruelty."

Chapter 8 Ravaging the Oceans and Seas

"Farming as we do it is hunting, and in the sea we act like barbarians."

—Jacques Cousteau

When I was growing up the cod was known as the "bread of the sea." It was one of the cheapest fish, so that when, as a treat, we had that very British delicacy, fish and chips, the fish was usually cod. We carried it back, wrapped in greaseproof paper inside sheets of newspaper to keep in the heat. In those days the cod fishing fleet hauled in huge catches. But gradually, as more and more people consumed more and more fish, the great shoals decreased in size. The price gradually went up. This led to increasingly strained relationships between Britain and Iceland.

Today the cod is a threatened species. So, too, are countless other fish. The depletion of the world's fish stocks resulting from the overfishing of our seas, lakes, and rivers, along with wasteful and unsustainable methods of extracting the catch, and the pollution of the water itself, is one of the most shocking ecological disasters of our time. Drift nets and long lines that stretch for hundreds of kilometers, nets with small mesh that catch young fish before they have a chance to mature, and the vacuum trawlers that suck everything movable into their giant maws, are some examples of unsustainable methods of fishing that destroy thousands of nontargeted species.

In 1993 there was an international ban on the vast drift nets that were threatening hundreds of nonfood species. Today their place has been taken by lines up to 130 kilometers long and with as many as 12,000 baited hooks for commercial fish. Before these sink under the water hundreds of thousands of seabirds, including endangered albatrosses and petrels, swoop down on the bait. Caught by the hooks, they are dragged under and drown.

If you've ever seen the torn and desecrated land desolate after the clear-cutting of a forest, you have a good idea of what the ocean bed looks like after a shrimp trawling operation. The trawlers cannot, of course, discriminate between shrimps and all the other ocean species that share the same habitat. Crabs, sponges, sea cucumbers, starfish, and countless other invertebrate creatures, all living peacefully among beds of sea grass and coral, are swept up indiscriminately, ravaging the foraging grounds of many fish species in the process. Years of trawling the ocean bottom has also destroyed coral communities and beds of sea grass that were foraging grounds for many species of fish, including the worms and other invertebrates that served as fish food. Dr. Russell Nelson, a consultant to the Gulf of Mexico Fishery Management Council and a veteran of over two decades of marine fisheries management and research, has written, "Where rocks and ancient drowned reefs have been eliminated, there is no possibility for recovery."

Though much yet needs to be done, there is hope. The amended United States Marine Mammal Protection Act permits the formation of special teams,"take reduction teams," charged with developing strategies to reduce the numbers of whales, dolphins, and porpoises in gillnets, long lines, and other mass-capture trawling methods. Working in coordination with environmentalists, marine scientists, animal welfare groups, and fisherman, these teams have come up with measures such as observational monitoring and closure of endangered areas. Additionally, fishermen may be required to put electronic pingers on their nets so that dolphins, seals, and so on are made aware of them. The penalties for violating these regulations are costly and fishermen generally comply. Certainly things have improved since the new rules were put in place.

In 1994 it was estimated that over 2,000 harbor porpoises in the Gulf of Maine were inadvertently caught in gillnets each year. Following implementation of the new regulations in 1999, only an estimated 270 were caught the following year. In the Mid-Atlantic, gillnets were blamed for killing an average of 350 harbor porpoises each year between 1995 and 1998. Again, after new regulations were introduced in 1999, the number killed the following year was fewer than fifty.

THE GREAT LAKES OF AFRICA

The great lakes of Africa are facing many problems. Often these begin when expatriates interfere in ancient fishing methods—in order to boost the catch—introducing programs that, while they may indeed result in immediate profits for the fishermen, also tend to lead to almost irreversible environmental damage. Particularly disastrous was the introduction of the large, succulent Nile perch into Lake Rudolf, in northern Kenya. These fish prospered, rapidly ate existing stocks, and then, starving, turned cannibal and consumed the young of their own species. Thus a lake that had fed thousands of people for at least hundreds of years was transformed into a watery wasteland in the space of just a few years.

I witnessed firsthand how changes in fishing methods

reduced the small sardine-sized dagaa of Lake Tanganyika from an abundant source of much needed food. In the early 1960s the sandy, pebbly beaches of the Gombe National Park were silver every morning with the little fish laid out to dry. In those days there was a fishing season—this corresponded with the dry season when it was almost certain that the fish would be dry by the end of the day, ready to be put in the sacks that would be sold not only in the nearest town, Kigoma, but sent by rail throughout the country. Large amounts also went to Zambia, where they provided protein for the workers of the great copper mines.

In 1961 Tanganyika became independent and, uniting with Zanzibar, became known as Tanzania. Some foolish decisions were made in the first flush of freedom from colonial rule: one such was allowing the dagaa fishermen to operate year-round. Again and again, during the long rainy season, the silver carpet laid out so hopefully after a night's fishing would become rotten and stinking after a day of rain. Mostly it was just left to decompose. Despite the extra demand placed on the fish, the nightly catch continued to be good until the Food and Agriculture Organization of the United Nations intervened. They introduced the fishermen to fishing with seine nets, and this would have a real impact on the wild stocks.

The traditional method of fishing, up until that time, was to lure the fish to the canoe, at night, with a lamp (at one time they had used wood fires), then plunge a net, rather like a giant butterfly net, on a long wooden handle, down into the water, scooping up as many fish as possible. When the seine net method was first introduced the fishermen were elated by the large catches. Several canoes were used to lay and draw in the net. They started far out and gradually pulled it in over a three- to four-hour period. But as time went on their excitement waned as catches became smaller. The mesh of the net was too small, and was destroying not only immature dagaa but also the young of other food fish.

By the 1990s the nets often returned almost empty. Finally, in 2000, the government banned the use of these nets.

HAZARDS OF FISH FARMING

Fish farming is an environmentally destructive, profit-driven industry that does to traditional commercial fishing what agribusiness does to family farming. Here's what happened when multinational aquaculture corporations moved into Canada, choosing the country because of its miles of coastline. The story has been documented, in chilling detail, by my friend the esteemed marine biologist and photographer Alexandra Morton in *Listening to Whales.*

For thousands of years the Chinese have raised herbivorous carp in freshwater ponds. There was no great overcrowding of fish, they were fed natural food, and the waste did not damage the environment. It wasn't until the Norwegians began modern fish farming, or aquaculture, in the early seventies that things began to go wrong. They were raising Atlantic salmon and, after starting them off in tanks, they put them in nylon net cages, *open to tidal waters*. This meant that wild marine life was exposed to the risk of disease from waste materials from the fish farms. This proved devastating to wild salmon everywhere the farms were established. Apart from this obvious flaw, however, Norway had established quite strict environmental restrictions—on size of net pens, number of fish, and so forth. This frustrated Norwegian fish farmers, who wanted bigger farms and more freedom. They went to Canada in the early 1980s because there they could do as they wished and began farming carnivorous *Atlantic* salmon in *Pacific* salmon habitat. They chose the Atlantic variety partly because they grow faster than Pacific salmon, and also because they are more docile so that more fish can be crowded in each pen. Concerns were raised about the domestic fish escaping and competing with wild stocks, but government scientists assured everyone there was no risk.

The first fish farmers in Canada, mostly Norwegians, at least consulted local people. But subsequently, as competition grew, new arrivals seemed not to care. In direct competition with the local fishing industry, the new fish farmers set their farms in areas favored by wild salmon. Thus in a place where hundreds of wild salmon had traditionally spent a few hours at a time, up to 150,000—later over 1 million—densely packed fish were placed in a pen that would be there for eighteen months before being harvested.

More and more fish farmers, Canadian as well as Norwegian, jumped on the bandwagon. Initially they were hailed as the saviors of remote coastal communities. But promised benefits did not materialize as mechanization steadily reduced the number of jobs. And, increasingly, local fishermen and scientists warned the government of the grave damage to local industry and to the environment that was taking place. Their warnings fell on deaf ears. Meanwhile, the local fishermen suffered. All the best fishing places were taken over by the floating cages, and commercial fishermen's catch dwindled even as the market price for it dropped. Stores were stocked with the cheaper products from the fish farms. Tugboat operators lost the places where they had tied up in storms, and yachters lost their best anchorages. A wild prawn fisherman operating not far from a newly introduced fish farm soon discovered that his best fishing ground went dead: His traps were empty and a black putrid sludge stuck to his traps. Marine mammals were found shot and acoustic deterrents installed to keep seals away from the cages also affected other creatures like whales and dolphins.

More and more people protested and eventually the government sent a team to investigate the complaints. This resulted in a map where certain key places along the coast were given a protected status. But when the fish farms moved into those places the government turned a blind eye.

The damage to the environment increased as time went on. The domestic salmon are fed pellets of fish meal, made by depleting wild stocks of small ocean-schooling fish. Doses of vitamins, minerals, and massive amounts of antibiotics were added to combat the sickness inevitable in intensively farmed animals. As a final insult these piscine prisoners were fed a pink dye. Wild salmon flesh is colored by the zooplankton they feed on in the open seas. If the dye wasn't added to farm salmon, a pale gray fillet of salmon would show up on your dinner plate.

In the farms, disease spread like wildfire, despite the massive doses of antibiotics given to the fish (there are more antibiotics per pound of livestock in fish farming than in any other form of farming). The fish were kept alive with ever more drugs. And then there was the waste. Industrial fish farms are the equivalent of floating hog or poultry farms. In the first year of the new millennium, the British Columbia fish farming industry dumped in the sea, *every day*, the equivalent amount of sewage that would be generated by a city of a million people. Around the cages the water went red with a bloom of algae that grows in water enriched by the nitrogen and phosphorus in the farm waste—tons of feces and uneaten food pellets. This bloom—which is everywhere where there are salmon farms—kills fish, and it numbs human lips. Yet the government shrugged it off.

We now know that there have been problems, sometimes major problems, wherever salmon farms are found. For example, there were proliferations of usually benign sea lice. Clouds of them leaked out of net pens—they then ate the skin off juvenile wild salmon and sea trout. Small numbers have virtually no adverse effect on wild fish, but the millions from the fish farms wreaked havoc and led to marked declines essential wild salmon stock. The sea lice have all but wiped out wild sea trout and salmon in Scotland, Norway, and Ireland.

In addition and despite assurances to the contrary, salmon from the Canadian farms escaped, passing every hurdle, inexorably displacing wild stocks of Pacific salmon. Even in Norway, where the regulations are much stricter than in Canada, about four million fish escape each year and in many rivers farmed salmon outnumber wild ones four to one.

HARMFUL TO HUMANS

It is not only the health of wild salmon and the livelihood of traditional commercial fishermen that are at stake. There are potentially serious health concerns for human consumers. In 2000, marine biologist Alexandra Morton cut open more than 800 domestic escapees that were caught by local British Columbia fishermen near her home in the Broughton Archipelago. She was conducting a study on the fate of escaped Atlantic salmon in the Pacific. Alexandra was concerned that the number of invading Atlantic salmon was being underreported and therefore the associated risks downplayed. She found in many cases that the uncooked flesh was so soft that it could be scooped out like mashed potato; some individuals had lumpy spleens, or orange-stained livers, or their vital organs melded together. She took swabs from two of the salmon that were covered with sores. She sent half of this material to the provincial government laboratory and half to a private laboratory. The results were very different. The private lab wrote "Every swab you took is crawling with bacteria. They're swarming the Petri dish." The bacterium was identified as *Serratia,* and was resistant to eleven of eighteen antibiotics. The government laboratory, by contrast, reported that *they could not find any bacteria at all!* (*Serratia* had recently been found on a salmon farm where sewage had leaked into the salmon pens in Scotland.) Nothing was done to further examine this health issue.

In the early 1990s, fish farms in Nanaimo, British Columbia, experienced outbreaks of furunculosis, a disease common in farmed salmon in Europe. It was introduced to Canada via a shipment of salmon eggs, which were imported *despite the advice of the government's own researcher that all imports should be banned.* Subsequently, a new strain of the disease swept through several fish farms and then appeared in wild salmon, killing over a quarter of the stock. Its high resistance to all three of the antibiotics approved by the fish farming industry strongly suggested its farm origin. But instead of ordering the fish to be destroyed, the government gave permission that they be treated with erythromycin—despite an earlier warning that fish so treated may not be used for human consumption. You might well have eaten some of those fish, as the farmers continued with business as usual.

Most of this information is not known to the supermarket shopper, who, naturally, prefers to purchase the cheaper, farmed salmon. The packaging in the farm-raised salmon also won't tell you that it's probably 50 percent higher in unhealthy fats than its wild cousin, and lower in desirable omega-3 fatty acids. And of course the label won't tell you that it probably contains high levels of a poisonous industrial flame retardant that's contaminated the food chain (PBDEs—polybrominated diphenyl ethers). Or that studies show that farm-raised salmon accumulate more dioxins and PCBs (polychlorinated biphenyls) that are linked to reduced sperm counts and cancer.

I have talked to many Native American and First Nation people. One and all are outraged by the fish farms. Right from the start the women told me that they would not cook the domestic fish, complaining that their flesh was soft, it was the wrong color and the wrong smell. Many of them, already, had seen a great reduction in the wild salmon, upon which many tribes count for food and business each year during the salmon run. The salmon rivers were becoming silted up as a result of clear-cutting, and poisoned because of the agricultural and sometimes industrial runoff. It seemed, to some of these people, that the fish farms were contaminating and infecting the last wild salmon and would bring about the end of their ancient tradition. We must not allow that to happen.

Alexandra told me that she has come to see the wild salmon as "a bloodstream, carrying essential nutrients up the sides of mountains to feed trees, bears, trout, indeed all the life around them. They are a gift the northeastern Pacific cannot live without and one that we will not be given again."

Today fish farmers in Scotland, who are anxious to avoid the mistakes made by the salmon farmers, are raising organic cod. "It's not every day an industry gets a second chance, and this is how we see it with the cod," said Karol Rzepkowski, director of Johnson Seafoods, one of the new cod farms. The Johnsons want to convince the world that they are farming their cod organically and sustainably. The cod's food comes from leftovers of fish already being caught for food in Britain. Cod do not get sea lice, so do not need treating. And they collaborated with the Royal Society for the Prevention of Cruelty to Animals to establish guidelines on welfare so that their cod should be "as content till death as a fish in a pen can be."

The Cod Wars

It began in 1956 when Iceland increased the fishing limits for British fishing vessels from four miles off her coast to twelve miles. Britain's fishermen protested but, despite government intervention, Iceland stood her ground, maintaining the need to conserve fish stocks. Eventually the two sides came to an agreement that limited the total amount of fish caught and provided some flexibility regarding the fishing limits. But after two years, in 1975, this agreement ended when Iceland declared a 200-mile offshore area out of bounds to British fishermen. This led to another phase of what had become known as the "Cod Wars."

At issue was the amount of fish deemed a sustainable harvest and the right to protect territorial waters. Hostile skirmishes between the opposing boats became frequent. Icelandic coast guard vessels approached British trawlers that trespassed into their territory and cut their nets. This led to some violent actions: Shots were fired and there were numerous rammings between Icelandic ships and British trawlers and frigates. No one was killed, but several ships were damaged during the conflict and some people were injured. After a particularly violent collision, it was thought advisable to consult the U.N. Security Council—which declined to take action.

After eight months of high tension a new agreement was finally reached. British fishermen were allowed within the 200-mile limit, but never more than twenty-four trawlers at any one time. And their catch was limited to 50,000 tons of cod annually. There were four conservation areas that were completely closed to all fishing, and Icelandic patrol vessels were allowed to stop and inspect British trawlers suspected of violating the agreement.

FARMING THE TIGER PRAWNS

The tragic saga of commercial prawn farming is just as shocking and, at the risk of seeming repetitious, I want to share this story also. The facts are taken from a report made by the Environmental Justice Foundation. At one time tiger prawns were a luxury only in expensive restaurants, but today they are cheap and available everywhere. What happened? It was in the 1990s that huge prawns began to appear on the market. Customers were prepared to pay more because they were so much less fiddly to prepare and, though expensive, were nevertheless cheaper than lobster. And then these prawns began to be imported in ever larger quantities: Annual sales in 2003 were some $50 to $60 *billion* worldwide, with most of the importers in the United States, Europe, and Japan. The market has been growing at about 9 percent per year.

Tiger prawns are the result of intensive farming—promoted to local governments by the World Bank as a way of generating income from exports, providing food for the hungry, and decreasing poverty in the developing world. They come from Ecuador, Honduras, Guatemala, and Mexico, as well as from Thailand, Vietnam, Indonesia, Pakistan, Bangladesh, and China. In some of these countries the government advertises on TV, promoting cheap loans for converting farmland to aquaculture. Believing they will become millionaires, many have mortgaged their land to build prawn farms.

It is not cheap. Prior to introducing the prawns, the ponds must be lined with plastic sheeting and then coated with a mixture of chemicals. Prawns are carnivorous and in order to raise prawns that grow quickly, the farmer must feed them protein—usually in the form of fish. And they must eat a very large quantity of fish to attain an acceptable giant size. Then, when the ponds have been stocked, with hundreds of prawns crowded together, it is of course necessary to put antibiotics in the water. Even so, the prawns, sooner or later, will almost certainly get sick. Many become deformed, covered with black marks—at which point more antibiotics are given. Even so, many die.

Not surprisingly it is only the wealthy who succeed,

since they can afford all the equipment, the chemicals, and the antibiotics, then buy another plot and start again when things go wrong. Many of the poorer farmers who take out loans from the World Bank to convert their farmland into tiger prawn aquaculture eventually find themselves financially ruined. In the west of Vietnam's Mekong, nearly half of the total number of prawn farmers lost all their money in one four-year period. During this same time, 70 percent of World Bank–financed ponds in seven Indonesian provinces were abandoned, and 50 percent in Thailand.

The impact on the environment has been devastating. Pesticides, antibiotics, disinfectants, and the uric acid excreted by the densely crowded prawns are pumped from the farms into rivers and the sea. Drinking water and agricultural land are affected. After being used for prawns the land cannot be used for growing rice or other agricultural crops. Thus when his prawn farm fails, a once independent farmer is ruined. In Bangladesh about half of the prawn farms are on land that once grew rice and thousands of subsistence farmers have seen their livelihoods destroyed.

Local fishermen suffer, too. The pollutants in river and sea, combined with the number of fish caught to feed the prawns, make it increasingly hard to make a living. Of particular concern is the large-scale destruction of mangroves that are cleared for the farms, for they are fertile breeding ground and habitat for many species of fish. It is estimated that about 40 percent of the world's mangrove loss is due to prawn farming, with devastating effect on local people. Eighty percent of the population in a coastal area of Ecuador, for example, have lost their main source of food as a result of the destruction of mangroves. In Thailand more than one-fifth of the farms in an area once covered by mangroves were abandoned after two to four years. When the prawn farming fails, the mangroves may never return and much of the pollution will persist. Thus the lives of local people will be affected long into the future. Already the loss of the mangrove barriers cost thousands of lives during the recent devastating tsunami.

Local farmers trying to make a traditional living by fishing or growing rice are naturally resentful of the prawn farmers, especially when multinational companies sweep in, clearing the forest, bulldozing agricultural land, establishing giant farms. People have lost their lives in at least eleven countries when gangs of poor, dispossessed, and desperate villagers attacked the prawn farmers and emptied their ponds. It has been necessary to install massive security measures, build tall fences, and install lighting.

Overall, then, prawn farming has led to debt and dispossession, illegal land seizures, abusive child labor (reported for some time by both the Save the Children Fund and Oxfam), violence, and horrible destruction and degradation of the environment. Yet despite all this, governments, anxious for a quick buck, persist in their efforts to expand the farms. Take Vietnam. In 2000 it was already the world's fifth largest producer: The prawn industry was bringing in $500 million per year. Yet the government wants to double capacity.

WILLING TO TAKE THE RISK?

There is one more negative. Western diners are, unknowingly, subjecting themselves to a serious health risk when they eat farm-raised prawns, for they are consuming a food whose size is utterly dependent on a heavy use of antibiotics and growth hormones. Already it has been shown that cancer-causing chloramphenicol and nitrofuran antibiotics were present in some prawns from China, Thailand, Vietnam, Pakistan, and Indonesia. It seems inevitable that other harmful substances will find their way to the dinner table.

OUR MERCURIAL OCEANS

There is a great deal more pollution of our seas and rivers than that resulting from fish and prawn farming. And much of this finds its way into the flesh of the seafood we eat. Mercury, a toxin found in many kinds of seafood, is associated with increases in blood pressure, impaired neurological function in infants, and reduced fertility in adults. The 2004 results from the Mercury Hair Sampling Project, sponsored by Greenpeace, revealed that 21 percent of women who were of childbearing age had mercury levels that exceeded the EPA's recommended limit.

In 2004 the FDA and EPA issued a joint advisory that consumers, but especially pregnant women, should avoid the following fish because they tend to store high levels of mercury: swordfish, shark, king mackerel, and tilefish. The government agencies recommended eating other seafoods that are lower in mercury, including shrimp, canned light tuna (such as white or albacore), catfish, and wild salmon.

WHAT YOU CAN DO

We have to face the fact that industrial fishing is causing terrible damage to our oceans, the wondrous ecosystem that sustains us all. Coral reefs—the rain forests of the sea—have declined 30 percent in the last thirty years, largely because of overfishing and shrimp trawling. Industrial fleets have fished out at least 90 percent of all large ocean predators—marlin, swordfish, shark, cod, halibut, skate, and flounder—in the past fifty years.

If you care about the sea and its wildlife, if you care about your health and that of your family, especially that of your children, and if you care about the livelihood of the commercial fishermen, there is, of course, something you can do. And you know exactly what that is! You can make ethically informed choices in the purchases you make from the store, or the food you order in a restaurant.

Refuse to Eat Farm-Raised Salmon

Demand wild salmon at your favorite restaurant. The wild salmon will cost a bit more—but even if this means you eat a little less, think how worthwhile your sacrifice will be! In fact, if you think about your own health and the health of the planet, it is no sacrifice at all. And incidentally, wild salmon tastes noticeably superior to farm-raised. Salmon lovers tell me that once you sample and compare the two you will never again eat farmed salmon. The pink salmon (also known as humpback), in particular, is always wild, it is abundant, and because it lives only two years and feeds low on the food chain it is one of the healthiest proteins available on earth.

Buy Organic Tiger Prawns

If, after reading about prawn farming, you are feeling that you might not enjoy them as much as before—or even that the thought of eating them, or anyone eating them, is distressing—take heart. If you have a craving for prawns, then search for the genuine organic tiger prawns exported from the warm coastal waters of Ecuador. Madagascar is working on a similar product. And you will be safe if you choose cold-water prawns—especially commended are those from Iceland.

Know the Facts About Seafood

But if overfishing is threatening our oceans and farm-raised seafood is often circumspect, how do we know which seafood to boycott and which kinds to support? Some of the best guides to eating environmentally healthy seafood are Monterey Bay Aquarium's *Seafood Watch Guides*. These free, pocket-sized guides help consumers discern what the best options are, and are published for different areas of the United States.

The Audubon and Wildlife Conservation Societies have also put out a *Seafood Wallet Card*. On it they tell the reader that consumer demand has driven some fish populations to their lowest levels ever. But, they say, "You can be part of the solution. You can choose seafoods from healthy, thriving fisheries. Which fish you buy at the market and [order from] the menu will determine the future of our oceans. You have the power to protect our marine life."

A lot of people carry these cards with them. My friend Tom Mangelsen found that his favorite seafood restaurant had shark and swordfish on the menu. He complained to the manager, who apologized. Tom was therefore shocked to find, when he next went there a few weeks later, that these endangered fish were still on the menu. He was angry and said that if he ever found shark and swordfish served there again he would cease patronizing the restaurant—and tell all his friends. It worked: That restaurant is now shark- and swordfish-free.

And it was public opinion in Taiwan that led to a government ban on shark fin soup at public banquets. Thus it is clear that we, the public, can have a major influence on the fishing industry, just as we did with dolphin-free tuna.

MONTEREY BAY AQUARIUM'S *SEAFOOD WATCH GUIDE*
SUSTAINABLE SEAFOOD GUIDELINES 2005

BEST CHOICES

Catfish (farmed)
Caviar (farmed)
Clams (farmed)
Crab: Dungeness
Crab: Snow (Canada)
Crab: Stone
Halibut: Pacific
Lobster: Spiny (US)
Mussels (farmed)
Oysters (farmed)
Salmon (wild-caught from Alaska)
Sardines
Shrimp (trap-caught)
Striped Bass (farmed)
Sturgeon (farmed)
Tilapia (farmed)
Trout: Rainbow (farmed)
Tuna: Albacore (troll/pole-caught)
Tuna: Bigeye (troll/pole-caught)
Tuna: Yellowfin (troll/pole-caught)

GOOD ALTERNATIVES

Clams (wild-caught)
Cod: Pacific
Crab: Blue
Crab: imitation/Surimi
Crab: King (Alaska)
Crab: Snow (US)
Flounder: Summer/Fluke
Lobster: American/Maine
Mahi mahi/Dolphinfish/Dorado
Oysters (wild-caught)
Pollock
Scallops: Bay
Scallops: Sea
Shrimp (US farmed or trawl-caught)
Soles (Pacific)
Squid
Swordfish* (US)
Tuna: Albacore* (longline-caught)
Tuna: Bigeye (longline-caught)
Tuna: Yellowfin (longline-caught)
Tuna: canned light
Tuna: canned white/Albacore*

AVOID

Caviar (wild-caught)
Chilean Seabass/Toothfish
Cod: Atlantic
Crab: King (imported)
Flounders (Atlantic) except Summer/Fluke
Groupers
Halibut: Atlantic
Monkfish
Orange Roughy
Rockfish (Pacific)
Salmon (farmed, including Atlantic)
Sharks*
Shrimp (imported farmed or trawl-caught)
Snapper: Red
Soles (Atlantic)
Sturgeon (imported wild-caught)
Swordfish* (imported)
Tuna: Bluefin

*Red asterisk indicates a FDA & EPA mercury advisory for women of child-bearing age and children.

Use This Guide to Make Choices for Healthy Oceans

Best Choices

These are your best seafood choices! These fish are abundant, well managed and fished or farmed in environmentally friendly ways.

Good Alternatives

These are good alternatives to the best choices column. However, there are some concerns with how they're fished or farmed—or with the health of their habitats due to other human impacts. Visit **www.seafoodwatch.org** to learn more.

Avoid

Avoid these products, at least for now. These fish come from sources that are overfished and/or fished or farmed in ways that harm other marine life or the environment.

Chapter 9 Becoming a Vegetarian

"Nothing will benefit human health and increase chances for survival of life on earth as much as the evolution to a vegetarian diet."

—Albert Einstein

I grew up, as mentioned, during the lean years of food shortages and rationing in England. Many people consumed whale meat (thank goodness no one in my family went that route) and a number of boardinghouses were caught serving their guests with meat that had been marked with a green stripe indicating it was "Unfit for Human Consumption." We had one "real" egg per person per week (mostly we used egg powder sent from Australia). We had a meat ration—I remember that very occasionally we had the traditional roast beef and Yorkshire pudding on Sundays—but we had tiny portions.

We were still living off food coupons when, in 1954, my mother, Vanne, and I went to stay with my aunt and uncle in Germany. We were utterly amazed to see the amount of food available in the defeated country—nothing was rationed and there was more in quantity and variety than I could remember. Uncle Michael was part of the occupying British army and one evening we all went out to dinner at a restaurant used by the upper echelons of the occupying force. We were early and the large dining room was almost empty. At least six waiters, immaculate in dinner jackets and bow ties, stood around the edge of the room. I was confused by the menu—there was so much to choose from and I was grateful when Uncle Michael ordered for me. It was spring chicken—I had no idea that half of a young hen would arrive, leg, wing, and all, on my plate. As I sat, wondering how on earth I could eat the thing, two of the waiters, seeing my confusion, came to my rescue. They whisked the plate away to cut it up for the poor *fraulein*—leaving me sitting with knife and fork poised for the assault! I was nineteen years old, and I was horribly embarrassed, but joined in the general laughter till we all had tears running down our cheeks.

I ate meat in those days. Chicken, steak, pork, bacon, fish—the whole lot. We all did. And it tasted especially good after the bleak war years. I knew that animals died. But I'd spent so much time with cows and pigs and hens, grazing and grunting and scratching about in the fields, and they seemed to have a happy life. Much happier than the Jews and Gypsies who had been tortured to death, better than the lives of the thousands of soldiers who had died in the war. And I assumed the farm animals had a quick and humane death.

Then, in the early 1970s, I learned about the horrors of intensive animal production as described in Chapters 5 and 6. This happened suddenly as a result of reading Peter Singer's *Animal Liberation*. I had never heard of a factory farm before and as I turned the pages I became increasingly incredulous, horrified, and angry. First, I learned about battery farms for hens. I knew something about hens. My first experience took place when I was just four and a half years old, an animal-loving little girl who lived in the middle of London. I can still remember the exquisite excitement I felt when we went for a holiday on a farm and I met cows

Mike and Sully were being sent to slaughter when they grew too big for the petting zoo. I wrote a letter to support the group of people protesting—now they live in peace at the Fauna Foundation sanctuary in Quebec. CREDIT: ROBERT SASSOR

As I travel around the world, many chefs and caterers offer to prepare organic meals. Linda Hampsten cooked this delicious repast for a small gathering of friends in Boulder, Colorado.
CREDIT: JEFF ORLOWSKI

Sheep grazing among the vines of Robert Sinskey's Farm, Napa Valley, California, keep the top cover trimmed and provide natural manure at the same time.
CREDIT: ROBERT SINSKEY

Many children today know little about the food they eat.
Some are not sure whether potatoes are picked
on the ground or in the trees!
CREDIT: TOM MANGELSEN/IMAGES OF NATURE

Obesity has become a leading cause of death.
CREDIT: ROY MCMAHON/CORBIS.

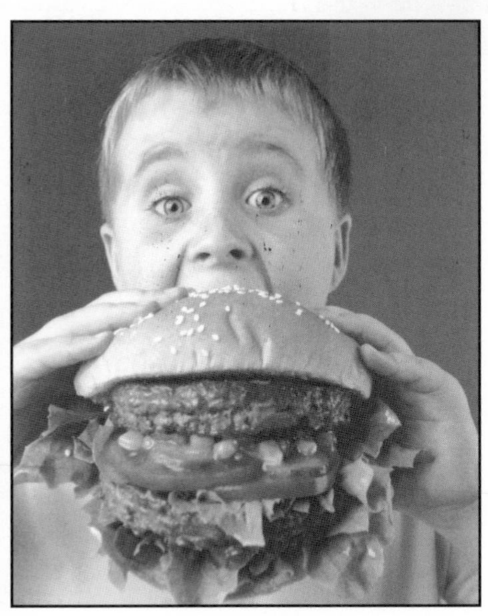

Is it surprising that there is an epidemic of obesity among our children? The growth of fast food, its usually unhealthy ingredients and the sheer size of the portions, is responsible for overweight children in more and more parts of the world. CREDIT: CHRIS EVERARD/GETTY IMAGES

One of hundreds of the sprawling center pivots that have jeopardized the ground water and aquifer in Nebraska. The arm, about a quarter of a mile long, rotates slowly around the pump in the center, releasing water in a huge circle.
CREDIT: TOM MANGELSEN/IMAGES OF NATURE

Forced to stand in mud and stinking liquid manure, these young cattle are curious about the photographer.
CREDIT: TOM MANGELSEN/IMAGES OF NATURE

Groups of women care for tree nurseries in the JGI TACARE (take care) program in Tanzania. We are gradually enabling these people to improve their lives: Now they are becoming our partners in our efforts to protect the Gombe chimpanzees.
CREDIT: KRISTIN MOSHER

and pigs and horses, close up, for the first time. And I helped to collect the hens' eggs each day. Most of them chose the nest boxes that were part of each little wooden henhouse (though some preferred the hedgerows).

I became curious. Where did the egg come out? I could see no opening big enough, and apparently I began pestering everyone with my question. With no satisfactory explanation I decided I would have to find out for myself. I hid in a stuffy henhouse, covered in straw, for so long that my family, who had no idea where I was, mounted a search party. It was my mother who saw me, rushing toward the house covered in straw, and full of the wonderful story of how a hen lays an egg!

About twenty years later I had the opportunity to get to know a hen and a rooster very well indeed. They were given to me and Vanne soon after we went to Gombe together in 1960. They were intended for the pot, but they were alive, with their legs tied—so they would stay fresh longer. Of course, I untied them, and they roamed our little camp, ridding it of all kinds of unwelcome guests—they simply loved scorpions! That was before I became a vegetarian, but any thought of killing and eating these engaging individuals—Hildebrand and Hilda as we called them—was out of the question.

Hilda was bold and demanding, insisting on her share of any food she saw me eating, and often wandering inquisitively into my tent. Hildebrand had a more nervous and retiring disposition—but he did develop a remarkably loud and raucous crow. This was okay in the morning—he was a great alarm clock, and I rose at 5:30 every day. But he also tended to start crowing at odd times, often making me jump. I found I was able to prevent this through what is known today as "behavior modification." Back then I just "trained" him! At first I threw a handful of straw or dust at him when he crowed in the middle of the day. After a while, I found I could predict, from small preparatory movements, when he was about to let forth his ear-splitting call—at which point I showered him. After a while I only needed to glare at him, and make a tiny hand movement, to prevent his territorial call. Poor Hildebrand—it must have been seriously upsetting to his masculine ego, and may have contributed to his shy personality! It was my time with those two engaging birds, knowing how they behave in freedom, that made me so deeply disturbed when I learned about the conditions under which they must exist in their minute prisons on a modern battery farm.

Next, in Singer's book, I read about pigs, and this not only made me angry—it made me weep. Perhaps I was being sentimental? Maybe I was influenced by reading *Charlotte's Web*, falling in love with that unforgettable literary character, Wilbur. Actually, I think it was the memory of a wonderful childhood holiday, when I was about eight years old, that upset me so when I thought of pigs in factory farms. On one of my walks I discovered a field of young saddleback pigs. I loved to watch them play, running and chasing one another, then resting and grooming each other in the shade. I used to go there with a picnic lunch every day—a skimpy wartime sandwich and a—probably wormy—apple. I tempted the pigs to come closer with my apple core, and one—I named him Grunter—eventually lost his fear. Not only did he accept my offering directly from my outstretched hand but—oh the excitement of it—he let me scratch behind his ears, along the stiff bristles of his back, under his chin. I was in heaven and no thought of his ultimate fate marred my pleasure.

I can still remember how I felt when I closed Singer's book. I thought about the delicious pork chops that I loved, the heavenly smell of frying bacon in the morning. And all the roast chicken, casseroled chicken, fried chicken, and chicken soup that I had enjoyed during my life. There was a sort of numbness in my mind. I knew I would not be able to keep from thinking about the images conjured up by the pages I had just read. When I saw meat on my plate, from that moment on, I should think of *pain—fear—death*. How horrible.

And so it was clear. I would eat no more meat. For another year or so I continued to eat fish. My son was catching fish where we lived in Dar es Salaam, Tanzania. And at least those fish lived free until their death. And, as he pointed out to me, if he didn't catch them, someone else surely would, what with the local fishermen, and the European Union and Japanese trawlers that were ravaging the seas off Tanzania. But the time came when the very feel of animal flesh in my mouth was so distasteful that I gave up fish as well.

Often I'm asked if I mind when people, with whom I am sharing a meal, order meat. I don't like it at all when they order veal, or great chunks of red meat (unless it is organic and free-range), but I believe that change must come from within, and I can usually find an opportunity to talk about why I am a vegetarian. I try not to do this until everyone has finished the beef or pork or chicken on their plates! Embarrassing and/or antagonizing people is no way to change their hearts—and after all, I myself am not a vegan. I never tell people not to eat meat. People don't like change. They don't like to be told what to do. No, my job is simply to explain the facts, very calmly, and hope that some will make changes to their diet. Sometimes it is quite spectacular: "I became a vegetarian last night," said one sixteen-year-old after he had listened to my talk.

Famous Vegetarians

Hank Aaron, baseball player
Benjamin Franklin, inventor, diplomat
Charles, Prince of Wales

Chelsea Clinton, daughter of President Bill Clinton and Senator Hillary Rodham
Leonardo da Vinci, artist, inventor, sculptor
Willem Dafoe, actor
Cameron Diaz, actor
Albert Einstein, scientist
Michael J. Fox, actor
Richard Gere, actor
Woody Harrelson, actor (vegan)
Steve Jobs, co-founder, Apple Computer
Ashley Judd, actor
Billie Jean King, tennis champion
Dennis Kucinich, member, U.S. House of Representatives
Mahatma Gandhi, peacemaker
Carl Lewis, track and field athlete
Tobey Maguire, actor
Demi Moore, actor
Edwin Moses, track and field athlete
Paul Newman, actor, philanthropist
Gwyneth Paltrow, actor
George Bernard Shaw, poet
Sofia, Queen of Greece
. . . the list goes on and on!

People also ask how I coped when I first gave up meat. Psychologically, I felt really good about it, especially during the first few months, when the smell of bacon still made my mouth water, and I could be proud of my strength of will! But soon I began to feel so much better physically, too. Lighter, somehow. Other people who have given up meat have said the same. Nor is it surprising, for when we eat meat we waste a lot of energy getting rid of the toxins in the flesh that the animal was also trying to get rid of before it died. Perhaps that was why I started to feel so much more energetic. Since 1986 I have been traveling three hundred days a year, lecturing, going to meetings, lobbying, teaching, and so on. Never in one place for more than three weeks consecutively, and usually only a few days. I honestly don't think I could have maintained this pace when I was thirty years old—and I believe that giving up meat is the reason why I can today.

If I was not on the road all the time I'd probably be a vegan. But it is not easy when you spend 300 days per year on the road and stay with people in all parts of the world to maintain a balanced diet without *any* animal products. It's fine if you can cook yourself, or go to a good vegan restaurant. But home-cooked vegan food and vegan restaurants are not options on much of my travels. So I still eat eggs and cheese, and I know that milk is present in many sauces and desserts. Whenever possible I get organic, free-range animal products, but it's often not possible.

Many people believe that meat is necessary for good health. The opposite is usually true. First, humans do not have the right kind of anatomy for frequent heavy meat eating. There is a difference in the length of the intestines of carnivores and herbivores. Carnivores have short intestines (about the length of their bodies) and are able to pass the nondigestible portion of their food quickly through the body before it starts to putrefy. Herbivores need more time to get the nutrients from the vegetable matter they eat and so have long intestines (about four times their body length). Humans have long intestines, too, so that flesh may sometimes stay for much too long in our guts. In other words, the human species does not have the physical attributes of the carnivore—and that includes ripping, slashing teeth and claws. And, finally, unless they only eat organic products, they are constantly contaminating their own bodies with the hormones and antibiotics fed to factory-farmed animals.

CONSIDER OUR CHILDREN

My great-nephew Alex was four years old when he discovered that meat was the flesh of animals who had been killed. He instantly decided that he did not want to eat any more of them. He proudly announced that he was a vegetarian (making the most of the difficult word in his four-year-old way—ve-ge-taaar-ian) when he went to elementary school, and he has kept it up for over a year. It was not easy for him at the start—he really enjoyed bacon, sausages, and a number of other little-boy foods. For several months he continued to eat fish. Then he was taken to an aquarium for the first time. He was transfixed by the tanks of colorful tropical fish. Soon he made the connection between these glowing creatures and his fish fingers and fish and chips. "I shan't eat pretty fish," he declared. But, after standing, enraptured, in front of tank after tank, he finally decided he would not eat *any* fish. Nor has he. Now that he is five, he has become a passionate opponent of the cruel practice of de-finning sharks for shark fin soup. He will, I suspect, be a staunch campaigner when he is a little older.

At no time did Alex criticize the meat-eating behavior of the rest of his family. But quite suddenly, three weeks ago, his young brother, Nickolai, who had been very carnivorous, often refusing all vegetables, asked why people kept hens. When he was told that it was for eggs and chicken he became agitated and announced that he, too, would become a vegetarian!

He has kept it up for two months now. Amazingly, the boy's father, who was always a big three-times-a-day meat eater, has become almost entirely vegetarian and only eats meat occasionally, usually out of the house. In the same way, my young friend Evelyn Kennedy, who also became a vegetarian when she was four years old, after seeing a truck full of sheep go by on their way to slaughter, gradually influenced the rest of her family to greatly reduce their meat intake.

I am always impressed when children, after suddenly realizing what meat actually is, quite refuse to eat it. Unfortunately their parents sometimes won't allow them to become vegetarians, wrongly believing that it will be detrimental to their health. In view of the fact that millions of the world's inhabitants never eat meat due to religious beliefs, and that many of them live to a truly venerable age, this is clearly nonsense.

Some parents worry about children becoming iron-deficient, since many people rely on animal products for their source of iron. But researchers have found that even vegan children (who don't eat any animal products, including eggs or dairy) can absorb enough iron if they eat plenty of vitamin C–rich plant foods along with a variety of beans, nuts, and seeds. Studies in the U.S. and U.K. also show that children raised on vegetarian and vegan diets appear just as healthy and normal as children who eat meat.

There is also a strong case for *encouraging* children to eat less meat. Food makes up 95 percent of our exposure to two highly dangerous carcinogens, dioxins and polychlorinated biphenyls (PCBs), with the highest levels coming from animal products, especially liver and oily fish. The recommended "safe limit" for intake of dioxins and PCBs keeps getting lowered, as researchers discover more and more health problems associated with these chemicals.

Shockingly, it seems that our children may be most vulnerable to PCB exposure while still in the womb. Developing fetuses that absorb PCBs tend to have more complications at birth, and more chance of behavioral and developmental deficits throughout their childhood. The same is true for dioxins—infants and children are the most vulnerable to exposure. And because dioxins also travel through the womb, nearly every baby born in an industrialized society has dioxins in its body. Dioxin exposure in the womb has been associated with numerous problems—birth defects, IQ deficits, attention deficit disorders, hyperactivity, and childhood depression.

The average daily intake of dioxins in the U.S. in 1999 was more than 200 times higher than the Environmental Protection Agency's cancer risk guideline. And one third of the British population, including toddlers and schoolchildren, may have dangerously high levels of dioxins in their diet, according to the U.K.'s Food Standards Agency. All this unsafe exposure is linked to increased respiratory diseases, and children getting more middle ear infections and allergic reactions.

Fortunately, we have a simple, everyday means to reduce the likelihood of exposure to dioxins or PCBs—stop or limit a child's intake of meat, fish, and dairy products. Joel Fuhrman, M.D., is an author and board-certified family physician who specializes in preventing and reversing disease through natural methods. In his excellent book on family nutrition and healthy appetizing recipes, *Disease-Proof Your Child*, he points out that we have been conditioned to believe that animal products are the best source of protein. In fact, peas, green vegetables, and beans have more protein per calorie than meat.

"Nobody can consume too little protein by eating less animal products and substituting more vegetables, beans, nuts, and seeds," Dr. Fuhrman states. The foods with the most nutrients per calorie are actually vegetables and beans. And the foods that are some of the most harmful are those packed with saturated fats (the fats we ingest from animal products). In the U.S. and other developed countries, overdependence on dairy products and meat has surely been a contributory factor to the current prevalence of heart disease and cancer.

In the spring of 2005 the BBC ran a story that seemed quite at odds with the above facts. The article quoted Professor Lindsay Allen as saying that parents who put their children on strict vegan diets could harm their development. She went so far as to say that denying growing children animal products in their diet was "unethical," arguing that animal source foods have some nutrients not found anywhere else and that pregnant women on vegan diets could be damaging their children while they were growing in the womb. As I had never heard this claim before, I was puzzled—until I realized that Professor Allen spoke as a representative of the U.S. Agricultural Research Service, an arm of the USDA, and that she was citing a study that was partially supported by the National Cattleman's Beef Association.

Children and Calcium

Many parents worry about growing children getting enough calcium to help with bone growth. While dairy products do offer calcium, physician Joel Fuhrman, author of *Disease-Proof Your Child,* explains that milk, cheese, and butter are especially high in saturated fats and that adequate sources of calcium can come from nondairy sources. Not only can parents find orange juice and soy milk that's fortified with calcium and vitamin D, they can also incorporate more plant-based sources of calcium into children's daily meals.

HOW EATING MEAT IMPACTS THE ENVIRONMENT

It is estimated that between one third and almost half of the world's harvest is fed to animals to fatten them for human food. In the United States, 56 percent of farmland is dedicated solely to the production of beef. In the U.K., about 70 percent of agricultural land is used for growing food for animals. So many animals are eaten in Europe, Japan, and other parts of the developed world that there is no way enough food can be grown in each country. The grass and grain needed to fatten the animals eaten as meat in Europe requires an area seven times the area of the European Union.

For this level of meat consumption to be maintained, European farmers must buy corn and other animal fodder from other countries. This is destroying the Brazilian rain forest, where vast tracts of virgin forest are destroyed each year not only to create pasture for cattle, but also for growing soy beans or corn, a great deal of which is shipped to Europe and Japan to feed their animals.

This means that, as local populations grow, land they urgently need for growing their own food is being taken over by foreign corporations. As a result, more and more developing countries are now, for the first time, dependent on imported grain. The situation will soon be particularly desperate for China as ever more people eat ever more meat and their own agricultural farming land is, at the same time, being cleared—and very rapidly cleared—for development.

Different Kinds of "Vegetarians"

Vegans are strict vegetarians, who avoid all foods of animal origin, including meat, poultry, fish, dairy products, and eggs. Many vegans even avoid wearing animal products, such as fur or leather.

Ovo-vegetarians eat eggs but they do not eat chicken meat.

Lacto-vegetarians eat dairy products in their diet, but not eggs. Some lacto-vegetarians refuse to eat cheese that's made with the coagulating agent rennet, because the usual source of rennet is the stomach of slaughtered newly born calves.

Lacto-ovo-vegetarians eat dairy products and eggs.

Pesco-vegetarians eat fish, and many also eat dairy products and eggs.

Semi-vegetarians, also known as flexitarians, will bend their own rules—eating poultry, fish, dairy products, or eggs—depending on specific social situations or health needs. But most draw the line at eating red meat.

FEEDING OUR APPETITE FOR MEAT

The process of feeding livestock with corn and soybeans grown on huge tracts of land, both within the country and overseas, is extraordinarily wasteful. The estimated amount of grain needed to produce different kinds of meat varies according to the vested interests of the organization producing the statistic. Thus the U.S. National Cattlemen's Beef Association claims that 4.5 kilograms of grain produce 1 kilogram of beef from a cow in a feedlot. But the U.S. Department of Agriculture's Economic Research Service says that the true amount is *16* kilograms of grain. The U.K. poultry industry claims triumphantly that 1.6 kilograms of feed will yield 1 kilogram of weight gain per animal—but the industry does not tell us that only 33.7 percent of each carcass is edible. So to create 1 kilogram of *edible* meat requires much more than 1.6 kilograms of feed.

The World Health Organization and the Food and Agriculture Organization of the United Nations have produced another kind of analysis, estimating the number of people that can be fed during one year from one hectare of farmland depending on what food is grown. These estimates range from twenty-two people per hectare for potatoes and nineteen per hectare for rice, to just one or two people for beef and lamb. Of course, people cannot live on rice or potatoes alone, but it clearly shows that the answer to feeding hungry people will not be found by increasing meat production—especially in view of the fact that even at present rates of animal consumption we are destroying the agricultural land of our planet. Instead, it is desperately important to change any culture that embraces a heavy meat diet.

Growing Fish Fillets in the Lab

When I saw an article in *New Scientist* about a group of scientists who were working on growing meat in a Petri dish I thought it was a joke. But it was a NASA-funded study, designed to see if the technology could be used to sustain astronauts on long space journeys. The team, led by Morris Benjaminson, professor at Touro College in New York, removed ten-centimeter chunks of live muscle tissue from freshly killed goldfish. This tissue was put into a cell culture fluid made from the blood of a cow fetus.

After a week the tissue had grown about 14 percent. So, says the article, "the team became the first in the annals of science to grow fish fillets in the lab." At a special press conference Benjaminson fried the fish bits with olive oil and herbs. No one was allowed to taste it, however, as lab-grown goldfish was not approved by the FDA for human consumption! And the cow serum could have been contaminated. But Benjaminson said he was delighted with the dish. "The fish smelled and looked like something straight out of the supermarket. The muscle we grew in vitro looked fresh. It looked pretty darn good. That is my humble opinion. The fish was fine—the serum was revolting."

Now he is planning to experiment with mushroom extract as a more palatable growth medium. But there are many problems before larger pieces of muscle tissue can be grown in the lab.

Whatever else, this is certainly an example of the ingenuity of science. Maybe this is the way forward for those who refuse to abandon meat. No more growing grain for animal food. No more degradation of the land or depletion of seas. Just huge warehouses with steaks growing in bins of mushroom juice! The cultures replicating themselves like yogurt!

IRRESPONSIBLE USE OF WATER

The relatively large amount of soybeans that can be produced per hectare comes with a cost. To produce one kilogram of soybeans 2,000 liters of water are needed, compared with about 1,900 liters for a kilo of rice. But 3,500 liters are used to produce a kilo of chicken, and for cattle a massive 100,000 liters is needed per kilo of beef.

In view of these facts, I believe that the single most important thing we can do, if we care about the future of the planet, is either to become vegetarians or to eat as little meat as possible, and that only from free-range organically raised animals. Indeed, WHO, FAO, and other organizations stress that in order to have a more healthy and sustainable human diet, we must reduce meat eating immediately with the goal of eating at least 15 percent less meat by the year 2020.

A CASE FOR EATING LESS MEAT AND SEAFOOD

The good news is that vegetarian—and vegan—food is not only good for the environment, the welfare of farm animals, and human health, but it is also delicious when properly cooked. I have shared many mouthwatering meals with my Hindu friends in Tanzania. And as more and more people are embracing a vegetarian diet, vegetarian cookbooks abound. Moreover, one is no longer regarded in restaurants as some kind of weirdo when one asks for vegetarian options. Indeed, all the best hotels, restaurants, and airlines offer a vegetarian choice. Even when there is nothing on the menu, chefs are becoming more inventive. Such was the chef in Sierra Leone at a little out-of-the-way restaurant who concocted a vegetarian dish so delicious that I went back to the kitchen to suggest that he add it to the menu. It is more difficult for me in Japan and many other Asian countries because fish is the basis of so many dishes. And it is even harder in some parts of the American Midwest where vast spare ribs and steaks are the typical fare. But even there I am no longer considered quite so peculiar.

I recognize that, for many people, giving up meat would be extremely hard. But if everyone knew and faced up to all the facts, most would either opt for drastically cutting their meat consumption and eating only free-range animals, or giving up meat altogether. For the mass production of meat on intensive farms is taking its toll not only, as we have seen, on the well-being of the animal victims, but also on human health. And it is wreaking havoc on the environment whether the animals are factory-farmed or grazed.

Jane's Diet

My son once said he wished someone would study me as he couldn't imagine how someone who "ate so little and of all the wrong things" could have so much energy!

So, what is my secret?

First, I became a vegetarian, which, as I mentioned, made me immediately feel lighter and more full of energy. Second, I live by my grandmother's dictate: You can eat anything you like—in moderation. Third, whenever possible, I eat organic food.

And now, for the curious, let me give you some idea of what I eat when I am at home in the U.K. or Tanzania:

Breakfast: Half a slice of whole wheat toast with Seville orange marmalade or marmite (which is very British, and thought of, by all my American friends, as a kind of edible creosote!) and a cup of coffee (shade-grown and organic and fair-trade, when possible).

Lunch: Broccoli or sprouts or some other vegetable, a small boiled potato—or half a jacket potato with cheese. Just occasionally this is varied by a homemade and very cheese-y macaroni and cheese or quiche. Then another cup of coffee and a couple of squares of chocolate or something sweet.

Supper: Scrambled egg on the other half of the breakfast toast and a glass of red wine. My idea of pure misery is large meals that leave me feeling full.

In between meals: Little snacks—a cookie, an apple, an orange—whatever happens to be around.

Supper is always preceded by a tot (shot) of whiskey—Scotch whiskey that is. No ice, just a splash of water. This has been traditional in our family for years. Wherever I was in the world I always raised a glass around 7:00 P.M., drinking a toast to my family and friends. Because I knew that my mother, Vanne, Olly (my aunt), and sister Judy would be doing the same. (My grandmother Danny was strictly a teetotaler except for brandy used medicinally: She always had a bottle in her wardrobe and it would last several years.)

During my endless tours, I cannot stick to the frugal meals I prefer. But I travel with the wherewithal to be independent when circumstances permit. I always have coffee, creamer, and sugar in my bags, along with packets of tomato soup. Then, with an immersion coil for heating water in a mug (or glass), I don't need to call on expensive (and often very slow) hotel room service if we arrive late. I developed the habit of squirreling away food during the many years in Tanzania when the economy was rock-bottom and it was almost impossible to buy bread, sugar, or other necessities in the shops. Then every uneaten roll, packet of sugar, and so on was taken back to the house. So today it is second-nature to take away a roll and butter not eaten on a plane and to save dinner rolls for breakfast. Not only does this give me independence, it also prevents the food from being wasted. And it saves money, too.

Chapter 10 Global Supermarkets

"The passive American consumer, sitting down to a meal of pre-prepared or fast food, confronts a platter covered with inert, anonymous substances that have been processed, dyed, breaded, sauced, gravied, ground, pulped, strained, blended, prettified, and sanitized beyond resemblance to any part of any creature that ever lived."

—WENDELL BERRY,
"THE PLEASURES OF EATING"

A generation ago there were probably about 800 items in the average U.S. supermarket. Today, we can find anywhere between 30,000 to 40,000 food items. We have packaged foods galore, and beyond that a bounty of fresh fruits and vegetables from every corner of the world. Some teachers are encouraging children to look at the labels as part of a geography lesson. It's now possible for supermarkets to stock almost any food from anywhere at any time. Of course all these choices seem like a boon to consumers who can afford them. Fresh broccoli every week of the year! Steamed asparagus whenever we have a yen! Fresh grapes in January!

What we don't often consider is the legacy of travel behind most of these foods. That broccoli we like to stir-fry every week may have traveled over 2,000 miles to reach our woks. The asparagus stalks could easily log 1,500 miles before they reach our steamers. The worldliest of travelers might well be midwinter grapes—whose trek is over 6,000 miles from a vineyard in Chile to a grocer in America.

THE MYTH OF "FRESH FOOD"

While the global supermarket represents a plethora of choices, it has also created an excess of unrealistic values and expectations. We have been hypnotized into believing that it is perfectly reasonable to walk into a supermarket and find any kind of food, from anywhere, anytime of the year. How different it would be if our produce came with a stamped passport showing every international and state border it passed through. As it stands now, few of us realize that a typical "fresh" food item in a North American household typically travels between 1,500 to 2,500 miles. Each year, food and agricultural products travel about 566 billion ton-miles within U.S. borders alone—not including imports from overseas—according to USDA estimates. Meanwhile our dependency on global food travel is only increasing. Our food travels 25 percent further than two decades ago and most of us don't even think about the ramifications. What it means is that it often takes far more energy to get our food to the kitchen table than the energy we receive from eating it. In fact, every calorie of energy we get from our typical high-travel supermarket food has already burned about ten calories of fossil fuel before it even reaches our mouth.

While it's hard to tally exactly how much global warming is affected by our travel-dependent system, we do know that nationwide trucking of food is responsible for a significant amount of fuel emissions. If food is flown into a country, the emissions are even higher. A "traditional" Sunday meal in Great Britain made from imported ingredients causes nearly 650 times more carbon dioxide emissions as the same meal made from locally grown ingredients. Let's also consider the amount of packaging waste that comes from processing and transporting food so that it can endure thousands of miles of travel. Many of our household processed foods are packaged with paper products (think lots of trees and paper mills) and plastic (which takes forever to biodegrade, litters the streets of every city, even blowing into the trees, and is overloading landfills around the world). Even our fresh foods are usually shrink-wrapped in plastic. Burning all this plastic isn't a solution either, since it fills the air with dioxins and other toxic residues. And because the growers, packagers, and grocers are never required to figure out what to do with or contribute any money toward the cost of its disposal, why should they bother to try to reduce the amount of waste they generate?

Imagine if our produce came with shipping papers showing the date it was picked and how many handlers it encountered along the way and everything that was done to keep it appearing appetizing. We might well discover that our "fresh" produce was actually picked a week or more ago and had numerous handlers—meaning numerous opportunities for bacterial exposure. We might find that it had been irradiated to keep it preserved during the long freight journey. Or we might realize it was picked prematurely and then "gassed" to ripen it after transport. Or spray-painted orange or injected with dye so that it appeared riper, more colorful. Or that it had been genetically modified so it would stay fresh for a longer stretch of time.

Food, Fuel, and Freeways

The Leopold Center for Sustainable Agriculture in Iowa compiled data from the U.S. Department of Agriculture to find out how far produce typically traveled to a Chicago terminal market. This chart compares that with the distance produce typically travels to the San Francisco Ferry Plaza Farmers Market, which specializes in locally grown food.

Chicago Terminal Market	San Francisco Ferry Plaza Farmers Market
(Averages)	(Averages)
Apples: 1,555 miles	Apples: 105 miles
Tomatoes: 1,369 miles	Tomatoes: 117 miles
Grapes: 2,143 miles	Grapes: 151 miles
Beans: 766 miles	Beans: 101 miles
Peaches: 1,674 miles	Peaches: 184 miles
Winter squash: 781 miles	Winter squash: 98 miles
Greens: 889 miles	Greens: 99 miles
Lettuce: 2,055 miles	Lettuce: 102 miles

As monoculture farms try to feed the global food chain, it's not surprising that they favor crops and animal products that travel well and stay fresh longer. Meanwhile, the world's cornucopia of delicious, organic, high-nutrition foods that are meant to be eaten ripe from field and garden are being excluded from the global market. One of the sorriest examples of this selective manipulation of crop species is the mealy, bland, and tasteless industrially grown tomato. This imposter may be forced to ripen in a refrigerated truck while traveling for a week before it reaches the supermarket shelf. Once there it may sit for another several days before it reaches your plate. And, of course, it may be a genetically modified variety. No wonder so many children are growing up with an aversion to tomatoes! They are probably as discriminating as animals, who instinctively avoid GMOs.

Perhaps all this transport and food manipulation would be justifiable if it were actually helping to feed regions of the world where fresh food is unavailable. But the fact is much of this food is shipped to communities and countries that grow the same foods in abundance. Returning to the mealy industrial tomato: It is quite likely to be offered for sale in a New Jersey supermarket in August while just a few miles away are some of the most sumptuous tomatoes in the world, grown by local farmers, sun-ripened and ready for the picking. But since the local tomatoes aren't part of the industrial food chain, they never make it to the supermarket shelves. As it stands now, many supermarket buyers are only set up to place bulk orders from warehouses that import food from packagers and factory farms all over the world. For a big chain supermarket buyer, there's little incentive to order food from a number of small local growers, when they can simply work with a few large-scale suppliers.

So not only has this travel-dependent system led to a loss in quality, it's also led to a tragic loss of small-scale growers. In the middle of October a supermarket in Massachusetts may sell a standard Red Delicious apple from Washington state along with a few other varieties that have journeyed from as far away as Japan and New Zealand. At the same time, the state's small-scale apple growers are struggling to make ends meet, unable to fit into the supermarket model and relying on customers to come to their roadside stands. Some of these family apple growers keep their heirloom orchards as expensive hobbies; others are forced to sell their ancestral heritage to developers.

Things used to be so different. Until the 1950s most grocers, all over the world, relied on fresh foods from nearby growers. Since then many changes have taken place, including the development of refrigerated trucks and, in the U.S., the interstate highways subsidized by the federal government that have enabled farms at one end of the country to connect with grocers on the other. Gradually, our local-based food supply network changed into a corporate-driven network that controls the majority of the growing, packaging, and shipping of U.S. food.

In their most recent book, *Hope's Edge*, Frances Moore Lappé and her daughter, Anna Lappé, point out that ten multinational food corporations control over half the world's food supply. In other words, a handful of CEOs are narrowing our grocery store selections to those products that provide the best economic returns to their corporations. Just as most people became resigned to the pesticides and genetically modified organisms in their food, so many are resigned to this corporate ownership that routinely sacrifices quality and diversity for the sake of mass production, and puts economic return above human and environmental health.

The organic movement has done a fine job of raising worldwide awareness and policing industrial farming. The next step is to create a more sustainable system. So don't be discouraged; we're on the right track, just read the next chapter to find out what more we can do.

A Spoonful of Sugar

To give an example of how ridiculous our global food distribution has become, California's Center for Urban Education About Sustainable Agriculture points to the saga of sugarcane.

Imagine you are sitting in a Hawaiian coffee shop pouring a packet of refined white sugar into your coffee mug. Would you know that the sugar you are about to drink was first processed in a plant across the street? But since the sugarcane was still in the pale brown raw stage, it was then shipped to the C&H (California and Hawaii) Refinery on the outskirts of San Francisco, turning it into snowy-white fine-grain sugar. Now it had to be packaged into little sealed paper packets for coffee shops, so the sugar then traveled across the U.S. to New York, where it was packaged and eventually distributed to restaurants around the nation, including the coffee shop in Hawaii. Ultimately, that package of sugar made a 10,000 mile round-trip processing journey before landing in your coffee mug.

Chapter 11 Taking Back Our Food

"Food is power . . . are you in control of yours?"

—JOHN JEAVONS

Go through the aisles of a local supermarket and you will see something almost unheard of five years ago—bins of organic produce, shelves stocked with organic ketchup and corn chips. From cans of organic heirloom beans to boxes of organic chocolate chip cookies—the demand is booming. In the U.S. it's known as organic, as it is in the U.K. In the rest of Europe it's called biofood. Whatever you call it, this growing trend is changing the course of global agriculture.

So what happened? What created this enormous increase in the production and availability of organic foods? One thing is for certain—it was not because the giant agribusinesses, chemical companies, and food corporations came to their senses and decided that organic farming was better for the environment and humanity. No, the reason that organic food products became more universally available was because people, the great general public, began coming to their senses. Gradually more and more people understood the dangers inherent in eating food contaminated with chemicals. They began worrying about their children's health. They began asking for organic products in supermarkets and grocery stores. They were even willing to pay extra for it.

What this exciting boom in organic foods has taught us is that we, the people, by virtue of the products we purchase—and those that we don't—can change global agricultural practices. It has become clear in this instance, as in many others, that while big corporations are rarely motivated to make changes as a result of ethical concerns, they are highly sensitive to how the public wants to spend its money. The dangling carrot is obvious—profit. And if people prefer organic over chemical-laden produce, then there is a big incentive to deliver the goods. The food industry is driven by consumer demand.

Unlike the many food fads that have come and gone, the demand for organic produce is unlikely to go away. Our desire for chemical-free food, an agricultural system that is in harmony with the environment, that will support farmers, and that will provide sustainable harvests in the developing world, is here for good. And it is growing bigger—one purchase, one bite, one vote, at a time. In 1990 consumers bought $1 billion worth of organic foods and beverages. Twelve years later that number went up to $11 billion. At this rate of growth, most of the food sold in the United States will be organic by the year 2020. And we see the same promising trend in Canada and Europe.

SENSIBLE FARMING

Another reason for the growing boom in organic foods is that, despite widespread belief that the system is inefficient and unprofitable, this is not so. Of course the big corporations tried to squash organic farming, claiming crop yields would be so puny and labor so intensive that farmers couldn't support themselves, much less feed the world. But this is being disproved again and again, as more and more farmers are discovering that organic farms with diverse, rotated crops are far less vulnerable to disease and more resilient in adverse weather conditions. In the U.S., the top 25 percent of sustainable farmers who are dedicated to creating healthy ecosystems without relying on chemicals now have higher yields than the nation's industrial farmers. The advantage of organic farming becomes even more striking during years of drought, when such farms produce much higher yields than farmers who grow with chemicals—as much as 33 to 41 percent more in 1998. And it is the same during flooding: When entire fields vanish under sheets of water, there is far more danger of soil erosion in monoculture fields than in those farmed organically.

On a separate note, organic farming could even help stop global warming. The Rodale Institute calculated that America could comply with the Kyoto Treaty's demand for a 7 percent reduction in greenhouse gases simply by making a full switch to organic farming. Industrial, monoculture agriculture uses around 30 percent more fossil fuel than organic.

It's not surprising, then, that more and more farmers are switching their crops to organic. In 1997 the U.S. had 1.2 million acres of land devoted to organic farming. Just four years later, in 2001, that number almost doubled, rising to 2.3 million. Even big businesses are switching to organic in the hopes of improving profit. Paramount, the largest citrus grower in California, is using sustainable methods on about one third of its acreage and nearly 300 acres produce organic food.

Indeed, all over the world farmers are switching to organic because it works better. As I mentioned in Chapter 3, the Ethiopian government is so impressed with organic farming's increased yields that it is adopting organic agriculture as one of its leading strategies for addressing famine and food security. Over 100,000 Mexican coffee growers have adopted full organic production methods and increased their yields by half. And if you scan the shelves of the grocers or supermarkets, you will see brands from many countries that are organic (as well as shade-grown and fair-trade, meaning the grower was paid a fair price rather than being exploited). The same is true for tea.

More Animals Agree: Organic Tastes Better

There has been no systematic testing of animals, but there are some observations that suggest that animals, with their often superior sense of smell and taste, choose organic rather than nonorganic when offered a choice. Keeper Niels Melchiorsen of the Copenhagen Zoo is quoted as saying: "For one reason or another the tapirs and chimpanzees are choosing organically grown bananas over the others." He noted also that when given an organic banana, the chimpanzees ate them skin and all. But if they were handed a nonorganic banana, the chimpanzees instinctively peeled the skin before eating it. Herbie, a chimpanzee at a sanctuary belonging to friends in Bend, Oregon, was offered four different kinds of food: tomato, eggplant, milk, and orange juice. In three out of the four (all but the eggplant), Herbie chose organic over commercially produced.

In fact, humans also prefer organic. For those of us who regularly eat organic food, having to eat nonorganic can be quite a shock for there is truly a difference in the taste.

DEEP ORGANIC, SHALLOW ORGANIC

The organic movement first began as an antidote to the growing corporate control over our food supply. From the beginning, the organic dream had three agendas: to grow healthy food in harmony with nature, to preserve the rich diversity of regional foods, and to create a new way of distributing food through farmers markets and food co-ops.

Many consumers choose organic foods because of the first part of the dream—the desire to eat safe and healthy food that is grown respectfully. As we make our organic purchases, we might also like to believe that we are supporting the thriving, diverse ecosystems these organic foods come from—the small family farms with rows of fragrant herbs and heirloom lettuces, stalks of corn, and mounds of zucchini plants.

Indeed, the organic industry's packaging—with pastoral scenes and soothing images of the natural world on many of the labels—certainly perpetuates this vision. In a 2003 Whole Foods Market survey the majority of people buying organic products said they believed their organic foods were grown on small farms. But the reality is not quite that idyllic.

Because organic foods are now part of a multibillion-dollar industry that's growing by 20 to 25 percent each year, is it any surprise that the corporate food giants are jumping on the bandwagon? For instance, organic lines such as Muir Glen and Cascadian Farm are actually owned by General Mills. The Heinz dynasty now owns an assortment of familiar organic labels, including Little Bear, Walnut Acres, and Health Valley. And Coca-Cola bought the hugely popular upstart juice maker Odwalla. With the recent development of organic high-fructose corn syrup, will it be much longer before we see Organic Coke?

But when the "organic" movement gets into the hands of big business we need to be wary. Look a bit closer at General Mills, the third biggest food conglomerate in North America, and you see that their principal stockholders include Monsanto, ExxonMobil, Chevron, DuPont, Dow Chemical, and McDonald's. Hain Celestial, another big name in organic foods, also has an interesting list of stockholders, including Monsanto, Wal-Mart, Philip Morris, and ExxonMobil.

The Once and Future Kings: Who Owns What

- General Mills owns Muir Glen and Cascadian Farm
- Heinz owns Hain, Breadshop, Arrowhead Mills, Garden of Eatin', Farm Foods, Imagine Rice (and Soy) Dream, Casbah, Health Valley, DeBoles, Nile Spice, Celestial Seasonings, Westbrae, Westsoy, Little Bear, Walnut Acres, Shari Ann's, Mountain Sun, and Millina's Finest
- M&M-Mars owns Seeds of Change
- Coca-Cola owns Odwalla
- Kellogg owns Kashi, Morningstar Farms, and Sunrise Organic
- Philip Morris/Kraft owns Boca Foods and Back to Nature
- Tyson owns Nature's Farm Organic
- ConAgra owns LightLife
- Dean owns White Wave Silk, Alta Dena, Horizon, and the Organic Cow of Vermont
- Unilever owns Ben & Jerry's

Many of these multinational corporate giants aren't contracting with the small-scale, sustainable farms we like to imagine. Instead, we see a new hybrid—the organic factory farm, with acres and acres of monocrops. For example, in California, where growing conditions are ideal almost all year-round, there are huge organic farms producing just one variety of carrot or vast fields of romaine lettuce that are eventually shipped cross-country in sealed plastic bags. These industrial organic farms are not only taking over farmlands, they are taking over the organic market's profits. Only 2 percent of California's organic farm operations (about twenty-seven large-scale growers) represent over half of the organic sales in the state.

Some see the difference between the large-scale industrial organic farms and small-scale, sustainable organic farms as the difference between shallow organic and deep organic. Shallow organic adopts the same success formula as corporate business—sameness, familiarity, and limited selections for long-distance hauling. They may not be using pesticides or GMOs, but they still rely on enormous amounts of subsidized water and are heavy users of fossil fuels. Because they insist on mass-producing monoculture crops, these growers are still trying to support an inherently weakened system. Instead of nourishing their soil with intensive composting and crop rotation, industrial organic growers purchase packaged bottles or bags of prefab "organic" fertilizers to spread on their crops—products that temporarily boost the soil but don't provide the viability of intensive organic farming.

The Tenets of Deep Organic

- Farmers are committed to biodiversity, raising different types of plants and animals, which are rotated around the fields to enrich the soil and help prevent disease and pest outbreaks.
- Resources such as water, soil, and air are respected and replenished, so the farm is a self-sustaining ecosystem and causes no harm to future generations.
- The waste within the farm's ecosystem doesn't contaminate the bordering land, air, or waterways.
- Agrochemicals are not used unless absolutely necessary (and only with great caution and the minimum application possible).
- Animals are treated humanely and are well cared for.

They are permitted to carry out their natural behaviors, such as grazing, rooting, or pecking, and are fed a natural diet appropriate for their species.
- Farmers are compensated fairly and workers are treated fairly and paid competitive wages and benefits. Farmworkers also work in a safe environment and are offered healthy living conditions and food.
- The farm contributes to the local food distribution, minimizing the cost, handling, packaging, and pollution caused by long-distance transportation.

Shallow organic still participates in an infrastructure that favors mass production, excessive packaging, and shipping food thousands of miles to the grocery store. Shallow organic still needs lots of middlemen and short-term fixes. And in the end, shallow organic's success is measured by profit, not sustainability.

On the other end of the spectrum, we have deep organic, the picture-poster farms that we like to imagine when we buy organic foods: small-scale, sustainable, and in service to the land, not in service to the "efficiencies" of corporate industry. Sustainable organic farmers don't mass-produce one crop, because they understand that the key to thriving ecosystems is diversity. Using crop rotation and composting, they mimic the intricate relationships of the natural world in order to create healthy soil and plants. Rather than importing water, they use natural resources to assure water flow. And when the crops are harvested, the deep organic farmer seeks out local outlets for sales, so that food is freshest, and pollution from travel and packaging is minimized.

It all comes down to different agendas. Deep organic tries to grow the most delicious, nutritious food possible while assuring the health of the planet. Shallow organic wants to meet certification standards within the industrial paradigm. It shouldn't be surprising, then, that corporations are pressuring government regulators to make the coveted "organic" label work better for industrial farming. In 2004 the USDA came close to relaxing organic certification standards so that GMOs could be used, fertilizers with sewage sludge would be allowed, and food could be irradiated for better long-term storage. The USDA also considered allowing farms to retain their organic certification even if they used animal growth hormones. Thanks to the outpouring of letters from outraged consumers and organic farmers, the USDA backed down.

This pressure isn't likely to go away as long as we have the profit-driven mandates of big business trying to squeeze into the restraints of organic certification. The more corporations we have vying for organic, the more they will want to shape it for their convenience.

It's Not Just Our Imaginations . . .

Organic Really Is More Nutritious

Over the last half-century, as chemical-based agribusiness became more prevalent, there's been a steady decline in the mineral level of fruits and vegetables. Many of us sensed that food grown in rich, fertile organic soil was more nutritious than conventionally grown foods, but there weren't many comprehensive studies that supported that hunch.

In 2001, nutritionist Virginia Worthington and the U.K. Soil Association both conducted extensive studies that compared the nutrition level of nonorganic foods to organic foods. Their findings validate the organic buyers' instincts. Compared to crops grown with chemical fertilizers and pesticides, they found that organically grown crops generally have better protein quality and higher vitamin C and mineral content—particularly calcium, magnesium, iron, and chromium. These differences were not so extreme as to base all your organic buying on higher nutrition. But it does confirm what common sense has told us all along—the healthier the soil, the healthier the food.

Although few consumers would question the advantages of having certification for nationwide organic standards, the USDA still has no regulations about the harmful wake of industrial agriculture. For instance, it doesn't mention how much subsidized water can be used. It does not regulate farm labor standards. It also doesn't limit how many fossil fuels can be used in production. Or monitor the waste caused by excess packaging. It does, however, demand the same amount of paperwork and documentation from large-scale farms as it does from small-scale operations. And because big producers have more resources to handle the paperwork and bureaucracy necessary to obtain USDA organic certification, and are also better equipped to meet the big orders for giant food processors and supermarkets, they are gradually forcing many of the pioneers who gave the label its marketability out of the picture. Thus, the bullying history of agribusiness repeats itself in the organic sector, as the small-scale organic family farmer gets muscled out by the industrial organic farmers.

THE COST OF ORGANIC FOODS

It's true, organic foods often come with a higher sticker price. Because of this, some argue that the organic movement is an elitist one—only offering healthy foods to the wealthiest. In fact, the mean income of a frequent organic food buyer in the U.S. is $43,280—solidly middle-class—and 31 percent of frequent organic buyers make under

$15,000 a year. And as more people buy organic, the more the prices will go down: Suppliers will have to place orders, which means more farmers will be encouraged to farm organically because they have a steady market for their goods. So this increases the easy availability of organic food, which often leads to a reduced price for consumers.

Some people willingly pay the price for organic foods, seeing this as a charity donation—a way to support the health of the planet or the farmers who are trying to do right by the land and their communities. In some spiritual communities it is even seen as "tithing"—which means taking a portion of one's income and giving it back to the world in a way that supports the greater good. Yet other people see it as a kind of health insurance payment, recognizing that by ridding their bodies, and the bodies of their children, of agricultural chemicals they may have fewer medical bills.

CAN WE REALLY AFFORD "CHEAP" FOOD?

There is another issue: We need to think carefully about the hidden costs of nonorganic foods. For years, we've been forced to buy into the chemical contamination of our planet and bodies under the guise of creating abundant, "cheap" food. But how cheap is it really? The real cost of industrial farming never shows up on the price sticker at the grocery store. The price never shows the money taxpayers pay for government subsidizing of agribusinesses. The grocery store price also doesn't reflect how much we pay for our damaged health and weakened immune systems. It's almost impossible to measure how much we spend trying to clean up and cope with the environmental damage caused by chemical intensive farming, but in the U.S., it's estimated to be $9 billion a year. We just cannot afford this "cheap" food much longer. About three million tons of pesticides, herbicides, and fungicides are used on this planet every year! Every continent is burdened with the ongoing cost of cleaning up this chemical contamination, especially in streams, rivers, and lakes.

WHAT YOU CAN DO

Unfortunately, it will take a long time for the earth to break down and heal itself from much of the poisoning caused by farming with chemicals. For now, we can't easily rid ourselves of what we have already done, but we can stop adding more. And there is one sure way to do this: Eat organic. If you buy "certified organic" food, you are guaranteed that it was grown without chemical pesticides, GMOs, chemical fertilizers, sewage sludge, or preserved with ionizing radiation.

Purists may think we shouldn't bother with organic certification if it's being co-opted by corporate interests. But we must also keep in mind that the booming organic certification movement is doing an enormous amount to protect consumers as well as the environment from the harms caused by industrial agriculture. Organic certification offers us important safeguards—forcing agribusiness to abandon some of its more destructive practices. A whole line of McOrganics or Organic Coke may seem like horrible aberrations, corrupting the sacred intentions of the organic pioneers. But any restraints we can put on the corporate world are important. And in some instances, corporations might actually give greater economic strength and consumer recognition to small, ethical businesses without corrupting their core values. For instance, when M&M-Mars bought Seeds of Change it helped finance a wonderful company that has remained dedicated to sustainable farming and protecting the purity and heritage of our seed supply.

As we said, the rising availability of organic foods gives us solid evidence of the power of consumers. It tells us that we are forcing the corporations, some of which are the biggest names in agribusiness, to change their farming methods in order to capitalize on the fastest-growing trend in the industry. And in the next chapter I'll tell you more about what you can do to directly support the small-scale, deep organic farmers we value so greatly.

Avoid Pesticides and Other Chemicals

Unless you primarily purchase organic foods, one item in three in your kitchen cabinet or refrigerator will likely contain pesticide residues. Government testing has found residues from as many as seven different pesticides on a single head of nonorganic lettuce. So if one of your main reasons for buying organic is to reduce your exposure to farm chemicals, you may want to substitute organic for the produce that typically has the highest levels of chemical residues.

Must-Have Organics

Consumer Reports and other public safety groups report that the following fruits and vegetables are especially high in chemical residues and, therefore, are the most important to buy organic:

- Raspberries
- Apples
- Peaches
- Cantaloupes
- Cherries
- Celery
- Green Beans
- Grapes and raisins
- Potatoes
- Spinach
- Tomatoes
- Winter squash
- Strawberries

U.S. shoppers should also avoid imported, nonorganic fruits and vegetables, since they consistently contain more residues than domestic samples. Also, many chemical companies take the pesticides that are banned in the United States and unload them in countries where there are fewer government restrictions.

Another bonus of certified organic foods is that they have to be processed without many of the food additives that have been linked to diseases such as cancer, heart disease, migraine, hyperactivity, and osteoporosis. For instance, the enhanced red color of some nonorganic strawberries comes from the fungicide captan, a probable human carcinogen that irritates skin and eyes, and is highly toxic to fish. Phosphoric acid in fizzy drinks has been linked to osteoporosis. Aspartame, an artificial sweetener in nonorganic foods, is linked to mood swings and migraines, and MSG (monosodium glutamate) is linked to asthma and headaches.

Feed Children and Babies Organics

As we said, children are especially vulnerable to the effects of pesticide residues. And we now have scientific evidence for what common sense already told us: Children and babies who are fed mostly organic foods have lower levels of pesticide residues in their bodies than children who are fed nonorganic foods. So it is well worth the investment in feeding children organic foods.

If you have no choice but to serve nonorganic foods, parents can reduce exposure by serving children vegetables with thicker skin, shells, or peels, since soft-skinned fruit and vegetables seem more likely to contain residues. According to research from *Consumer Reports*, pesticide residues in a single serving of peaches "consistently exceeded" the EPA's safe daily limit for a forty-four-pound child. Since pesticide residues can be transferred (often in a more concentrated form) through the placenta and breast milk, it is especially important that women who are pregnant or breast-feeding follow these guidelines.

The Growing Popularity of Organic Wines

Until recently, most vineyards have been treated to frequent and liberal doses of chemicals, massively contaminating the surrounding environment. But as people's awareness of vineyard chemicals increases, organic wine is becoming ever more popular. Recently I discovered there's been a sudden surge in the popularity of organic wine in China, of all places. And, as one might expect, it is taking off in California. In the early 1990s the Gallo Wine Company in Sonoma County shifted 6,000 acres of wine grapes from industrial growing methods to organic. Gallo found they could produce the same yields as they did with chemical farming—and at a lower cost per acre. Fetzer, America's sixth-largest wine producer, has converted all of its California vineyards into organic production.

At home in Bournemouth, my sister Judy and I always buy organic wine when we can. I recently received a brochure describing the Robert Sinskey Vineyards in Los Carneros, Napa Valley, California, in which I read that a flock of 687 sheep were turned loose in the vineyards. In a time of unusually heavy rains the cover crops and weeds grow quickly, and it is hard to get equipment in to manage them. But the sheep can wander, nibbling down the tops of the cover crops. At the same time the sheep manure fertilizes the soil, enhancing mycorrhizal fungi and bacterial action in the soil, which increases the vine's absorption of certain vital nutrients as much as fivefold.

The sheep help to create a healthy environment where nutrients break down slowly—as nature intended—and the vines get these nutrients as they need them, thus fortifying their immune systems, and picking up trace elements that are missed when they are force-fed with synthetic chemicals. And, of course, the flavor of the grapes is improved.

In June 2005 I was given a bottle of wine called "Gorilla" from a 100 percent organic vineyard, Comte Cathare, in France. The vintner, Robert Eden (son of Sir Anthony Eden, the onetime prime minister of Britain), sprays his crops with nettle tea—a wonderful natural pesticide. In fact, he goes to great lengths to preserve a respectful relationship with the earth, conserving energy whenever possible and rejecting any use of chemicals. He is getting a team of horses for plowing his fields so that no gasoline or diesel will contaminate the land. And the new wine cellar is being made from straw bales. They also get natural power from solar panels and a small windmill. He even follows a biodynamic calendar, taking into account the position of the sun, moon, and planets before planting, harvesting, cultivating, or pruning.

Very soon Comte Cathare will introduce a new wine on the market—with a percentage of the sales donated to help the chimpanzees. This can be a wonderful campaign that will help our brand-new Jane Goodall Institute in France, help the chimpanzees, and help the sale of wines produced in an ethical way.

Chapter 12 Protecting Our Family Farmers

"The best fertilizer is the footprint of the farmer."

—ANONYMOUS

Thanks to growing awareness among consumers, and the fact that more and more are buying organic food, there is a chance for small family farms to survive. And some of them, despite difficulties, are determined to do so. One farmer in Virginia has recently gained fame for his deep organic approach. Joel Salatin understands that growing food is a natural, not a mechanical, phenomenon, and he works hard to ensure that his farm reflects the diversity and the cycle of life that is best for the land he cultivates. He is a master of what he calls "pig-aeration," where corn and hay are scattered regularly among the dung that naturally collects from the cattle he raises. When the cows are let out to pasture, pigs are let in and naturally root through the dung for buried kernels of corn. The result is naturally fermented and aerated fertilizer—the perfect formula to replenish the hay fields, from which he then feeds his cattle.

But farmers like Joel need all our help if they are to survive. As he recently noted in a *New York Times Magazine* article, "Not only is there not government support for what we do, there is a profound antagonism at every level for what we do." On top of this, consumers, for the most part, are increasingly attracted to the prepackaged or fast food that relies on consolidated agricultural networks, networks that promote conformity and shareholder profit margins. Meanwhile, farmers like Joel are doing their best to make a different kind of investment—an investment in the future of the land they cultivate, and the health of those who eat their food.

LOCAL: THE NEW ORGANIC

Fortunately for Joel, and the growing numbers of farmers who share his values, quietly, slowly, and profoundly a food revolution is taking place. This is usually known as the "local foods" movement (some call it the "new organic"). And this movement is offering hope to the ethical family farmers across the country. It is a beautiful and timely convergence that speaks to all our concerns about industrial agriculture.

Environmentally conscious consumers know that eating local foods represents an opportunity not only to support small-scale farmers who are respectful, deep organic stewards of the earth, but also to help reduce the pollution caused by excessive food travel and packaging.

Health-conscious consumers find that eating local fresh foods from sustainable farms minimizes the pesticide residues, antibiotics, growth hormones, and hidden GMOs present in so much industrial farm products. Eating from local sustainable sources also offers a better diet: a diversity of nutrition-rich, fresh foods that will enable them to minimize their consumption of packaged, processed, or fast foods with their high sugar and fat content.

Politically minded consumers who are increasingly alarmed by the homogenization of global culture—with corporate chain stores taking over so much commerce in our cities, towns, and villages—find that eating local foods is a vote against imperialism and a tangible way to fight for a return to independent ownership. Eating locally also helps preserve the heritage and traditions surrounding food and cultural identity that a fast food world threatens to destroy. Food enthusiasts simply find delight in the quality and array of delicious choices that a noncorporate, local food supply offers.

And everyone seems to appreciate the renewed relationship with their food supply that the local food movement offers. When you think about it, the act of eating—putting something directly into our bodies—is an intimate process, and it is only natural that we should desire a more intimate or at least traceable knowledge of the people, land, and waters that provide our food. There was a time when we could claim this knowledge. Now we have a hard time even naming the continent many foods came from. Eating locally is a way to regain this sense of community and connection that was lost when big business came between us and our local food supply.

When we push a cart through the supermarket—with its aisles of jet-lagged produce, packaged foods, harsh fluorescent lights, and noisy computers scanning bar codes—shopping feels like a chore. But when we visit a farmers market with its rainbow of colors, smells, and flavors, shopping is transformed into a delightful outing.

It is the same the world over. The markets in the developing world are especially wonderful because so much is new. There the produce is laid out on the ground on woven mats or displayed on wooden tables. There are little piles of carefully counted fruits and vegetables, many of them well known and familiar, like the bright yellow or green bananas, orange-red tomatoes, white cloves of garlic, and the less colorful potatoes, onions, cabbages, and so on. But there are always some that are strange-looking and exotic with unknown names. I can never forget the first time I saw watermelons at a market in Aden where the ship stopped on my first ever voyage to Africa. There was a dark-skinned little boy standing there, exquisitely beautiful. Next to him were glossy dark green rinds cut open to expose the glorious bright red of the flesh from which the black seeds gleamed like ensnared ebony beads. (That was in 1957 when fruits from distant lands were not seen in the U.K.: I had never seen a watermelon.)

Mostly there is a wonderful, distinctive smell about the African, Asian, and Latin American street markets—a mixture of delicate fruit, earthy vegetables, and tangy spices. Very different was the market in Jakarta at the height of the durian fruit season. The smell of this giant fruit, when ripe, has been compared to that of a blocked drain, which is a fair description! Most Westerners are so put off by the smell that they cannot bear to even try the fruit. Which is a big mistake, for it is absolutely delicious. No wonder it is so much loved by the orangutans when they find it in the forest.

The farmers markets of the United States and Europe are similar, places where local producers set up their own stalls, showcasing and selling foods that they personally grew, reared, caught, brewed, pickled, baked, smoked, or processed. Most stalls offer tastes of homemade jam or artisan cheeses, slices of just-picked apricots or nibbles of tender lettuce. The beauty and vibrancy of local treasures even attract tourists, who find farmers markets one of the best places to literally get a taste of a region. The arts and crafts of local cottage industries are often displayed alongside the treasures of the land. And because farmers markets cater to small-scale producers, they are one of the best retail outlets for the small-scale, sustainable family farms that so many of us want to support when we buy organic products at the grocery store.

Considering the growing popularity of local foods, it's not surprising that these vibrant farmers markets are becoming increasingly successful, often linking urban or suburban areas with regional small-scale growers. In 1994 the number of farmers markets officially registered with state governments in the U.S. was 1,755—by 2004 the number jumped to 3,706. Sales of organic food in the U.K. increased tenfold over the last decade—from just over £100 million in 1993–94 to £1.12 billion in 2003–04. And there are probably thousands more unregistered small-scale farmers markets throughout both these countries.

With its stalls of organic produce alongside stalls of fresh-baked breads, medicinal herbs, wildflower honey, and an assortment of crafts by local artisans, the Olympia, Washington, farmers market is typical of many local farmers markets. One of the more regular fixtures at this thriving waterfront market is the Boistfort Valley Farm stand. When we think of the difference between shallow and deep organic, Boistfort Valley Farm is about as deep as it gets.

Relying on a variety of traditions, such as crop rotation and cover cropping, owners Mike and Heidi Peroni are dedicated to growing what they call "life-sustaining" food for their community in a manner that supports the soil, air, and water but also looks after the region's wildlife. For instance, they are especially concerned not to allow the farm to have any environmental impact on the nearby Chehalis River where wild salmon spawn. Their fields are surrounded by hedgerows that feed foraging elk and deer, while also serving as protective barriers for their crops. Mike and Heidi maintain patches of snowberry and crab apple trees throughout the farmland to provide resting places for the songbirds. And they even grow random stalks of sunflowers so that bees and other insects can move through the farm with ease and birds can eat the seeds.

Beyond selling produce through farmers markets, the Peronis also have about 200 Community Sponsored Agriculture (CSA) members. For $500 a growing season, customers receive twenty weekly boxes of Boistfort Valley Farm produce—enough fresh produce to last most four-person households a week. Every box contains a selection of cooking vegetables as well as an assortment of herbs, salad vegetables, and fruits. The Peronis also contract with organic fruit growers in eastern Washington, so they can provide CSA members with orchard delights, such as cherries, peaches, nectarines, apricots, and apples.

Many of their CSA customers liken the weekly boxes of organic goodies to receiving a Christmas package every week—with some households even battling over who gets to pick up the order and unpack the surprises. Because the Peronis grow vegetables specifically for the Northwest climate, rather than for shipping or shelf life, customers discover foods that would never be found in a long-haul refrigeration truck, much less the produce section of a typical supermarket. For those who might be baffled by the Shunkyo radishes, Kabocha squash, or Roma beans in their weekly box, Mike provides inspiring recipes for the week's selection, as do many farmers who want to help CSA families learn to cook with local, seasonal foods.

"I love the idea that every day I'm in the field, I can put faces on the people who are going to eat what I pick," says Mike. "And at the end of the workday, there's nothing better than sitting down at the kitchen table, knowing that in that moment 200 families are probably sharing the same meal that we grew."

Sample CSA Box of Weekly Produce from Boistfort Valley Farm

Spring	Summer	Fall
1 bunch carrots	1 1/2 lbs Roma beans	1/2 lb Chanterelle mushrooms
1 bunch golden beets	2 sweet onions	1 Kabocha squash
2/3 lb snow peas	2 Italian eggplant	6 ears corn
1 1/2 lb shell peas	1 bunch carrots	2 lb green beans
1 Winter Density lettuce	2 oz basil	1 pint Sungold tomatoes
1 bunch Mizuna mustard	1 bunch Tuscano kale	1 green leaf lettuce
1 bunch arugula	1 head celery	1 bunch Italian parsley
2 pints strawberries	2 lb summer squash	2 lb plums
1 bunch Shunkyo radishes	1 Butterhead lettuce	1 lemon cucumber
1/2 lb broccoli	2 lb nectarines	1 bunch summer savory
1 bunch Joi Choi	1 bouquet sunflowers	fresh lilies
garlic tops		
fresh lavender bouquet		

The original idea for a more direct farmer-consumer partnership program began in Japan where it was appropriately called *teikei*, literally meaning "putting the farmer's face on the food." The idea was born in the 1970s when a group of Japanese women decided they were fed up with unhealthy processed foods grown with heavy doses of chemicals and pesticides. But just as importantly they wanted to know and protect their country's endangered rural farmers.

This wonderful, commonsense idea of eliminating chemicals, processing, and middlemen and creating a more direct relationship with rural farmers eventually took hold in the U.S. when the first Community Sponsored Agricultural program was formed in 1985. As the interest in eating more local, sustainable foods has grown, so has the CSA movement.

Nowadays there are more than 1,000 CSAs throughout the country, reaching consumers in rural outposts as well as suburbanites and city dwellers.

Most CSA programs are similar to the one at Boistfort Farm. Consumers pay farmers to provide them with boxes of weekly, seasonal produce, fresh from the local farm. In a sense, consumers become shareholders in the farm, but instead of looking after maximum profit, these shareholders measure their returns in terms of freshness, nutrition, connection with farmers, and peace of mind. For many, the lure is fresher, tastier, and often organic produce. Some like the idea of directly supporting local farmers who are responsibly looking after the region's environment. But almost everyone—farmers and consumers alike—say the main benefit is the relationship with the grower.

My Visit to Cabbage Hill Farm

On a fine spring day toward the end of 2004 I visited Jerome and Nancy Kohlberg at their Cabbage Hill Farm in Mount Kisco, New York. This farm is entirely organic, and we had a delightful morning visiting their animals—heritage breeds of pigs, cattle, sheep, and poultry. For me, it was like a return to a farm of my childhood. All the animals were grazing or resting in pastures that nestled among the woods. We spent some time with their Large Black pigs—they came to the fence to be admired when called by name. And we were introduced to the Devon cattle, Shetland sheep, Maran chickens—and a prize group of Shetland geese, a very rare species, rescued and brought to the farm to breed. We also visited the greenhouse where trout and tilapia are raised, and shown the sophisticated and extremely efficient water system that has been set up.

As we arrived back at the house the rain clouds overhead suddenly opened and it began to pour—a real deluge. And this was accompanied by the most exhilarating thunderstorm. Just as we got up to have lunch a blinding streak of lightning, almost overhead, fused all the lights. It became very dark indeed—pitch-dark in the loo!

I was really impressed by my experience on the farm and I asked if they could send me a description for this book—it illustrates so many of the things we believe in so passionately. It follows:

"We are dedicated to the preservation of small farms, heritage breeds, sustainable agriculture, and biodiversity. We believe it is important to preserve pure gene pools, and that heritage breeds are the best animals for small farmers because they need little care: They breed and birth naturally, they roam and graze on pastures, and do not need antibiotics or hormones or chemically fertilized fields. Compost and manure are returned to the fields to enrich the soil. Our fish are raised in large tanks, and are fed organic food. The water from the tanks is pumped through floating beds of greens that absorb the nutrients from the water and then returned to the fish tanks clean and pure—there is no wasted water in this setup.

"All the produce from the farm is sold to local restaurants, including the Flying Pig in Mount Kisco, which we set up to show how a small farm can provide fresh and naturally grown produce to a small community, which, in turn, can help sustain a small farm.

"We believe that the patenting by the big chemical companies which ties the farmer to buying seed, chemical fertilizer, and weed control will destroy farmers all over the world."

POUND, DOLLAR, PESO, RUPEE: LOCAL IS AFFORDABLE

Perhaps the reason the local foods movement is also called "food democracy" is because it represents a way to take back control of our food supply. For consumers, direct sale opportunities, such as CSAs and farmers markets, usually offer healthier food at more reasonable prices than we find at grocery stores. Grassroots efforts for direct sales of local foods also help bring fresh produce to low-income households, with cities such as New York allowing residents to use food stamps at farmers markets or to pay CSA fees. Many CSA programs also offer installment plans and reduced fees.

Small-scale family farmers often prefer selling their products directly to the public through farmers markets, CSAs, and other outlets, such as farm stands, u-picks, and food co-ops. Generally speaking, the closer the food stays to where it was grown, the greater the percentage of the sales goes to the farmers, as well as to the regions that support them. A study by the U.K.'s New Economics Foundation shows that whatever the currency you use—pound, dollar, peso, or rupee—when you spend it on local foods you generate twice as much income for your community as you would by buying the same food from a supermarket.

FARMING IN OUR CITIES

Brian Halweil in *Eat Here: Reclaiming Homegrown Pleasures in a Global Supermarket* provides a comprehensive overview of modern farming within cities, estimating that some 800 million people on every continent are engaged in urban agriculture, mostly for their own families. Jac Smit, who heads the Urban Agriculture Network, which collects information about modern farming in cities around the world, believes that urban farming, which began springing up in the 1970s in the cities of Latin America, Asia, and Africa, will become increasingly important as more and more people migrate from rural areas. Already the huge Asian cities, such as Beijing and Shanghai and Jakarta, face horrendous traffic problems predicted to worsen, much of it caused by the ten-ton trucks distributing food. I know these cities and it is a nightmare already! Clearly if they can produce increasing amounts of fresh food from local urban farms it will help.

I became excited when I read Halweil's descriptions of successful urban farming projects. He estimates that, worldwide, about one third of the residents of big cities get their fresh food from urban agriculture—growing vegetables and fruit in backyards and basements, vacant spaces, and rooftop gardens. For example, more than half of the five million residents of St. Petersburg in Russia grow food in the city. Ten percent of greater London is farmed, with more than 30,000 allotment gardeners—including some 1,000 beekeepers. And the number of community gardens in Toronto increased from fifty to 120 during the past ten years.

Earth Pledge, an organization in New York, is actively promoting rooftop gardens. It hopes that the greening of the city's roofs will not only provide fresh produce, but lower the air temperature, help prevent pollution, and retain storm water. They have a splendid organic kitchen garden on the roof of their headquarters in Manhattan. In Mexico roof gardens have been designed with hydroponic technology. In Morocco there are roof gardens where vegetables are planted in old tires filled with compost: Their yield is thought to be as good as that of rural farms. Moreover, water use is reduced by 90 percent, since it is collected and recycled as it drains through the bottom of the beds. And many of these projects are providing meaningful work for the very poor—for example, in an Argentina slum where people pick organic materials from garbage dumps and make compost—for their own gardens or to sell.

In poor urban areas, in both the industrial and developing world, families typically have to spend more than half their income on food, since they cannot buy in bulk. And often the areas where they live are underserved. In the Anacostia section of Washington, D.C., the people were without a single supermarket for years, just fast food outlets and convenience stores. When a farmers market, with produce from urban gardens, was established it gave them their first reliable source of fresh food in years. In Havana, the U.S. embargo, followed by the Soviet collapse, forced the government to step in to help the residents, and today about 90 percent of the city's fresh produce is grown in its own urban farms and gardens.

Urban farms protect watersheds that supply city water and they enable people to reestablish their contact with the land as well as grow their own food. They also provide a use for food waste that would otherwise end up in landfills. And, very importantly, they provide meaningful work. Halweil describes the imaginative project of Wally Satzewich, onetime cabdriver, and his wife, Gail. Between them they manage twenty residential garden plots in Saskatoon. They either pay rent or barter a food basket in lieu of rent. They farm organically, frequently rotating crops to minimize damage by insect pests. They have built up a twenty-member CSA and they also supply some of the best known restaurants. Moreover, they make enough money to live comfortably: From one of his plots he earned $3,900 in one season. Wally is enthusiastic, sharing his project on his Web site. How hopeful it feels, thinking of all these people, all over the world, rolling up their sleeves and reconnecting with the soil. It makes me smile.

SAVING FAMILY FARMS

Every time we buy locally grown food we are supporting the beleaguered small family farmers, like Joel Salatin and Mike and Heidi Peroni. Each year more than 8,000 square miles of U.S. land is swallowed up by the vast suburban sprawl of strip malls, convenience stores, and housing developments—land that is often ideal for agriculture. Farmlands that don't get developed are often taken over by large corporate farms. Conservationists worldwide are starting to believe that preserving sustainable, local farms is almost as important as protecting our wild lands.

Of course the obvious benefit of preserving family farms is that they give us access to diverse, nutritious, and regional foods. But local sustainable farms also support the healthy characteristics of our communities. In rural areas where the family farms are protected and still producing, we often find higher employment and more thriving local businesses, schools, parks, churches, and community organizations, according to the Institute for Food and Development Policy. But when there is a dearth of local farms, rural communities tend to die off or struggle to survive.

Brian Halweil, a leading voice in the local foods movement in addition to being the author of *Eat Here,* speaks of rural "food deserts" where isolated households have not only lost their community services, they don't even have access to grocery stores or farmers markets because corporate grocers don't find it profitable to serve these poor outposts, and local farms have been lost to industrialization. Many of these households wind up feeding their families the highly processed food at convenience stores placed beside highway gas stations.

WHAT YOU CAN DO

There are many ways in which individuals, or groups of individuals, can make a difference.

Save Land

Recently I came across a story about a group of enlightened people in the town of Scrabble, in West Virginia, and their efforts to save their vanishing farmlands. As the town is only a one-and-a-half-hour drive from Washington, D.C., the farmland is prime real estate—perfect for developers.

The group began its efforts when a local 300-acre farm was offered for sale. It would be purchased by a developer and turned into two-acre plots, without respect or appreciation for the community and environmental assets around which they were developing. As a result, town residents banded together and cashed their stocks and retirement accounts, and mortgaged their homes in an effort to buy the farm themselves.

In a few weeks they had raised $700,000 but, although this was a remarkable feat, it was still less than the amount offered by the developer. However, although they did not secure that particular farm for their town, they are not giving up. They are now prepared and, with a financial network in place, are ready to fight to save other local farms as they are put on the market. And then, with the kind of options for farmers described in this chapter, the farms may actually be revitalized and help feed the residents of the town with locally grown, healthy food.

Buy from Local Farmers

Somewhere near you there are family farmers who are trying to do right by the earth—who are trying to feed their families as well as their community with integrity and respect. Aside from eating less meat, buying your food from a local farmer who is a good steward of the earth is one of the most effective contributions you can make to the health of the planet. The more we invest in these farms the more we build the world we want to live in. And the more likely we are to pass on the kind of world our children and grandchildren deserve to inherit.

Shop at Farmers Markets

Farmers markets are one of the most ideal sources of local, sustainable foods. You'll discover that almost all freshly picked vegetables from local farms taste phenomenally better than produce that's traveled thousands of miles before reaching your local supermarket. Be sure to ask stall operators what's the best and freshest produce. Farmers markets can be found in cities all across the U.S. The USDA has a complete listing of markets in all fifty states at their Web site (see Resources).

Become a Shareholder in a Farm

A share in a CSA costs about $300 to $500 for a twenty-four- to twenty-six-week growing season. In return, you receive weekly boxes of fresh seasonal fruits and vegetables. Many CSA programs accept monthly payments, and you may be able to buy a half-share rather than a whole share. (See the Resources section for a Web site that can help you locate a CSA near you.)

Join a Food Co-op

A food cooperative is a member-owned business that provides groceries and other products to its members, usually at a discount. Many of the products lining the shelves of co-ops are organic and much of the produce comes from local family farms. To find a co-op near you, check out Web sites such as Cooperative Grocer (www.cooperativegrocer.coop) and Local Harvest (www.localharvest.org). Joining a co-op is usually easy, typically requiring you to pay some dues.

Farmers markets, CSAs, and co-ops are probably the purest expression of the original vision of the organic movement because they offer direct contact with local growers and also offer direct support of their efforts. Detractors say that the commitment to eating local, sustainable foods is nothing more than a pie-in-the-sky utopian dream. But Joan Gussow, author of *This Organic Life* and one of the people who helped write the USDA organic certification standards, puts it this way: "I've often been told my vision for a food supply that is mostly based on local farmers markets and CSAs is completely unrealistic. But I believe that the current practice of food distribution is even more unrealistic, considering that we're running out of petroleum, and we will eventually be unable to ship our food all over the place."

Chapter 13 Eat Local, Eat Seasonal

"It's difficult to think anything but pleasant thoughts while eating a home-grown tomato."

—LEWIS GRIZZARD

The whole idea of eating local foods becomes especially alluring if we think in terms of *seasonal* foods. Start by refusing the tasteless, bloated, and artificially dyed grocery store strawberries that have been picked far away—unripe so as to travel better—in anticipation of the mouthwatering local ones that will be ripe late June and early July. They will be especially delectable if they are organic and free of the chemical pesticides and fertilizers that concentrate in the flesh of thin-skinned fruits like strawberries. I can no longer eat any other kind of strawberry.

When I was young we had no option but to eat those things that were in season. We waited eagerly for the first green beans or the first brussels sprouts, vegetables that came just once a year. We couldn't grow much in the sandy soil of Bournemouth, but we did grow runner beans and some other vegetables. In the summer my grandmother Danny made gooseberry fool, rhubarb and apple pies, and blackberry and apple pies with custard, and blackcurrant jelly. Uncle Eric used to make chutney with green tomatoes—his own special secret recipe. In the autumn Danny bottled fruit and made jam. We could never lay out apples to dry because of the salt sea air of Bournemouth, but I used to taste them when I went to stay with friends. They became more and more wrinkled but sweeter as autumn gradually changed to the long, dark days of winter. Today, fewer and fewer people lay in stock for winter days—for now foods are packaged and shipped from all around the globe so that customers can buy just what they want, just when they want it. Only when we decide to return to eating more local foods shall we regain our appreciation for the gifts of the different seasons and once again live in harmony with nature's cycles. Then, even as we grieve the last, now tough and stringy runner beans, we shall be eagerly anticipating the first of autumn's plums that are soon to be ready for the picking.

Nutritionist Joan Dye Gussow, an esteemed pioneer in the local foods movement, tells the story of walking out of her front gate one snowy February day in her upstate New York home. On the ground, peeking through the snow, she saw a bright orange-red object. The color was so out of place in the winter landscape that she leaned down for a closer look. There she found a nectarine with one bite taken out of it. "I thought of how it probably came from a foreign country," she said, "and was picked unripe, then put into a refrigerated shipment, and ripened artificially as it was hauled all the way up to New York state where someone bought it at a supermarket, bit into it, and of course finding it tasteless, mushy, and disgusting, just thoughtlessly tossed it away. In that one nectarine I saw all the waste and senselessness of our travel-dependent food supply." If only that person had simply waited until local fruits ripened on the trees—there is nothing quite like biting into the sweet flesh of a fresh, utterly ripe nectarine. So succulent that the juice is sure to run down your chin.

PROTECTING OUR HEIRLOOMS

When we commit to eating more local foods, we not only help the farmers of our region, we protect our indigenous foods and animals. What we need is a world of adjoining regions each known and valued for the special and distinctive produce that has been uniquely flavored by the particular soil and climate where it grew. But if we do not fight to protect these unique regions, with their beautiful orchards, nut trees, and traditional crops, they will all too soon be wiped out by housing developments and strip malls or reduced to monoculture farms that mass-produce corn and soy for processed foods and animal feed.

Alarmed by the fast food and supermarket homogenization that threatened to take over the world's food supply, Italian Carlo Petrini founded the Slow Food movement in 1986. When we think of the Slow Food movement, we might imagine a global mission to protect our regional markets—which means actively supporting crop biodiversity and sustainable agriculture as well as the regional food knowledge and customs that come with a locally based food supply. What began as a small, grassroots, Italy-based organization of a few concerned citizens has evolved into a rapidly expanding worldwide organization of over 80,000 members in fifty countries, with 12,500 official members in the U.S.

One of their many international efforts is to create an "Ark," a catalogue of farm animals, crop species, and agricultural techniques that are in danger of extinction. A quick study of the Slow Food Ark shows that the United States has many national treasures at risk. For instance, the Ark lists California's Blenheim apricot, which is described as "tart, honeyed and intensely aromatic." Biting into the deep golden flesh is said to be an "unforgettable experience"—and we agree—yet it may soon be forgotten unless their orchards are protected from developers.

Other fruits are endangered because they just don't fit the heavy travel supermarket model. New York's yellow-toned Spitzenberg apple, with its subtle streaks of red in the flesh, is perfect for eating fresh from the tree, but it does not ship well and is thus in decline. We may also lose the Rhode Island Greening, an excellent cider and pie-making apple that doesn't fit the generic, supermarket apple model. The Shagbark hickory nut of Wisconsin is endangered due to shrinking rural areas and the fact that manually recovering the meat is laborious. The nut has a delicate, sweet taste with no bitterness and once upon a time it was a traditional food item during the region's winter holidays. Fortunately, there are a few elders who still value the nut and sell it at local farmers markets in Madison.

When we protect the diversity of crops and food traditions of a region, we also preserve the special knowledge of how to cultivate and care for them. If we allow our local farms and orchards to be paved over or carved into housing developments, we also lose the growers who understand the soil, who understand the weather, who know how to work in harmony with their environment—knowledge that is far more valuable than chemical mixtures or instructions from genetically engineered seed manufacturers.

AS THOSE WHO CAME BEFORE

When we value the foods of the seasons, we help protect priceless ancestral wisdom as well as the heirloom seeds that have been passed on and perfected through many years and many harvests. Hundreds of years ago Iroquois white corn provided daily sustenance for the six nations of the Iroquois based in New York state, Pennsylvania, southern Ontario, and Quebec. Legend has it that the Iroquois gave this corn to George Washington's troops so they could survive winter at Valley Forge. As it is with many foods that are crucial to native survival, the Iroquois white corn was also integral to many of their spiritual ceremonies.

To preserve its delicious earthy flavor and varied texture, the Iroquois farmers developed special methods of growing the corn, orally passing down the ancient knowledge from parent to child, generation to generation. The small number of growers who still cultivate the corn know the ancient secrets for keeping the corn pure and delicious, and they also know how to time the planting and harvest to protect it from being contaminated by a neighbor's commercial corn. The corn is still roasted or hulled and milled in a log cabin on the Iroquois Cattaraugus Reservation in western New York state.

TRADITIONAL FOODS ARE THE HEALTHIEST

The diverse foods of the world's regions also serve another purpose—besides offering us biodiversity and wonderful flavors, researchers now believe that a diet based on traditional, indigenous foods is often healthier for us. For instance, East Africa's Masai people are cattle herding pastoralists, with a diet that's primarily based on milk and meat, meaning 66 percent of their calories typically come from fat, primarily saturated fats. To put that into context, most North American dietitians recommend no more than 30 percent of our daily calories come from fats. Timothy Johns, a Canadian ethnobotanist at McGill University, studied the Masai diet and lifestyle and discovered they also ate a variety of wild, indigenous plants that are high in antioxidants and other properties that tend to reduce cholesterol levels. His research concluded that balancing the diet with indigenous plant foods is one of the main reasons the Masai people typically have no health problems related to the consumption of so much saturated fat.

The people of the Tohono O'odham tribe of southern Arizona are trying to bring back their traditional wild foods to help address their soaring heath problems. Prior to 1960 diabetes was unknown on the Tohono O'odham reservation. Since then, the tribe began to adopt the typical North American diet with an emphasis on saturated animal fats, processed foods, and a lot of sugar. By 2004 the tribe had one of the highest rates of adult-onset Type II diabetes in the world, with 50 percent of all tribal adults diagnosed, and even some children being diagnosed with diet-related diabetes.

Researchers eventually concluded that the tribe's traditional, indigenous diet of wild foods—such as tepary beans, mesquite beans, cholla (cactus) buds, and chia seeds—helps regulate blood sugar and can actually reduce the incidence and impact of diabetes. So with the help of a USDA food security grant, the Tohono O'odham Community Association began sponsoring outings in the Sonoran Desert to collect the tribe's wild foods. They also distributed more than 1,000 packets of traditional seeds. Now gardeners on the reservation are restoring the traditional foods as well as the wild ones. Ultimately, it is hoped that the tribe will dedicate more of its 10,000 acres of cultivated land to traditional food crops, reducing the amount of land that is now used for cash crops such as cotton and hay.

When the tribe moved away from its indigenous diet, it wasn't just the people's health that suffered. They also experienced a decline in those cultural practices that were associated with traditional foods. With this renewed commitment to eating wild foods, the youth are once again learning the customs surrounding harvest. The tribe even brought back the rain dance ceremony, which had not been performed for thirty-five years. Now when the tribal members dance beside the new crops of traditional foods, asking for rain to sustain them, there is renewed hope for their harvest as well as their tribe.

Wendell Berry, the well-known Kentucky-based author who often writes about the ethic of contemporary agriculture landscape, laments the loss of so much agrarian knowledge—wisdom that took generations of family farmers to acquire. Imagine the fate of the Tohono O'odham people if the knowledge of traditional foods and customs had died off with the tribal elders?

SLOW FOOD PICKS UP THE PACE

The Slow Food movement recently found momentum at a conference, Terra Madre, held in Italy early in 2005. There 5,000 farmers from 130 countries met to discuss the future of food and how the individual farmer—and the local delicacies they produce—can survive amid the commercialization that is so dishonestly promoted by governments and multinational agribusiness.

This gathering was in part a response to meetings of the WTO, the World Trade Organization, and other organizations that seek to determine how foods are produced and traded. It was attended by a diverse cross section of international food producers, including rice farmers from Benin, beekeepers from Azerbaijan, Maori potato producers from New Zealand, and Vermont cheese makers from the U.S. It served as a powerful reminder that determined farmers can survive amid the globalization of today's marketplace. And by banding together, these small Slow Food producers are succeeding in raising the profile of the movement and strengthening their markets. Most importantly, they are finding a collective voice, having realized that individual voices are all too often silenced by the power of international corporations and the political parties they support.

But there is strength in numbers. As one conference attendee said, "It's so good to know we're not alone."

CHANGING THE WORLD ONE HARD HAT AT A TIME

"Hope cuisine, not haute cuisine."
—THE FARMERS DINER MOTTO

Barre, Vermont, is a small, working-class town and the center of the state's granite industry—a place known for quarries and tombstone carving. Here on Main Street, in a small fifteen-foot-wide building that has been feeding residents for over seventy years, is one of the most revolutionary and progressive businesses in the nation—the Farmers Diner. In this tiny diner with its green vinyl booths and white Formica countertops, we find proof that a more hopeful harvest is already under way.

The Farmers Diner is the brainchild of owner and organic farmer Tod Murphy, who one day wondered if it would be possible to create a restaurant with good old American diner food at good old affordable prices but have most of the food come from local and sustainable sources. He took over the landmark diner in 2002 and since then has turned it into a prototype for a national franchise that may someday put McDonald's and Starbucks to the test. More than a meal, the Farmers Diner represents the cutting edge of the local foods movement. So far, the Farmers Diner is spending 70 cents out of every dollar on products from farmers and small-scale producers who live and work within seventy miles of Barre. You'll find milk shakes, burgers, and omelets on the menu, but the burgers and milk will come from local grass-fed cows, the omelets from the eggs of free-range and small-flock hens. And if the season is right, you're likely to find a few surprises, such as a slice of green zebra heirloom tomato under your burger bun or a dish of local organic strawberry ice cream.

The way Tod Murphy sees it, he's building a bridge between the sustainable local foods movement and the everyday customer who may not be able to afford a meal at Alice Waters's Chez Panisse in California, but would still appreciate the riches of local, seasonal fare. (In fact, the diner's tomatoes are grown by one of the farmers who used to supply organic tomatoes for Chez Panisse.) Many of the residents who first started coming to the diner in 2002 didn't really care much about local sustainable foods, they just wanted good, reasonably priced diner fare. While customers have come to appreciate the local connection, it is the food that keeps them coming back.

Murphy speaks fondly of his regulars, such as the elder Vermont ladies who come from the assisted living retirement housing two blocks down from the diner. "They almost wept when they tasted our sauerkraut," he said. "It's not cooked, it's just chopped cabbage with sea salt, and for a lot of these ladies over seventy this is the sauerkraut they remember from when they were little girls." They have the same reaction to the diner's beet salad made of four different varieties of local, organic beets. "For many elder customers, the resurgence in local fresh food gives them a chance to remember and retaste the food from childhood, before industry got in the way," said Murphy. "And hopefully the desire and the commitment to pay what is needed to taste it again."

Sometimes it's not until customers start reading the menu and place mats, profiling all the local farmers who brought them their breakfast, lunch, or dinners, that they are even aware that the diner offers local riches. The customer literature explains that the diner only sells animal products that have been raised without antibiotics and hormones, and all livestock has access to the outdoors and pasture as a principal part of their feed ration, thus explaining why the butter is bright yellow. "That's the way butter is when the cow is outside eating fresh grass, not chained up in a barn," said Murphy. The literature also gives the history of the diner's hamburger, how it made a seventy-mile route from farm to plate instead of the almost 2,000-mile journey of most industrial beef.

Tod Murphy may be an organic farmer with an idealistic vision, but he is also a savvy entrepreneur. It certainly hasn't escaped his attention that just about every food and travel magazine now includes a favorable feature about local foods. He also noted that food trends, such as gourmet coffee, seem to first appeal to consumers in the top upper 10 percent income bracket, but eventually become popular among the middle class. He sees the local foods movement eventually becoming a mainstream mandate.

His vision is to create a national franchise of Farmers Diners—with about five to ten diners working with a central pod of local suppliers. So far his sights are on a pod in the Pioneer Valley of western Massachusetts, another one in the Hudson Valley/New York City area, and eventually moving on to the San Francisco Bay area and Portland, Oregon. According to Murphy, each pod will help eighty to 200 local farms become secure in their economic viability. For now he's seeking investors while he continues to infiltrate the working class of Vermont. He likes to recount the time a crew of construction workers came into the diner for some hamburgers. They had come to Barre from another part of the state to work on a local gas station.

"One day at a quarter to twelve in walks a whole crew of construction guys looking for burgers," Murphy recalled. "They have Dallas Cowboys stickers on their hard hats, and they immediately take over two booths with their cement-covered boots sticking out in the aisles."

When the burgers were served, one of the guys declared, "This is the best hamburger I've had since I was a kid in

Montana. This has got to be Black Angus from Colorado or Montana or Wyoming."

But his friend said, "What the hell are you talking about? Can't you read the menu? It's from right around here—in Starksboro, Vermont."

The other guy got mad. "No way did this come from Vermont. I'm telling you I grew up in the West and this is Western beef." For a while it looked like a fight might break out.

This story especially delights Murphy. "These guys didn't come into the diner because we offered local food. They came in to get hamburgers and get back to the job. But suddenly they are having a discussion about local food. You now have a guy sitting there thinking holy smoke this tastes different and better than anything else because it's local and you got another guy who's now got emotionally involved in defending the position that this is local. This is what the victory for our side really looks like."

Another encouraging trend is Burgerville, a chain of thirty-nine fast food restaurants in the Pacific Northwest that was dubbed "America's Freshest Fast Food" by *Gourmet* magazine. Their menu is nearly identical to McDonald's, but the mini-chain buys the bulk of its ingredients from farmers in Oregon and Washington. Some of its big sellers are local Tillamook dairy products. This may not be remarkable for a restaurant that features local sustainable foods—but what is remarkable is that McDonald's is catching on. Drive by the Seattle McDonald's that serves the Space Needle tourists, and you'll see a brightly lit sign, boldly advertising its exciting new menu item—local Tillamook ice cream.

WHAT YOU CAN DO

Obviously unless or until we see dramatic changes in our farming and grocery infrastructure, it will be hard for many, if not most, of us to get all our food from local sources. Especially if we live in one of the so-called food deserts discussed in Chapter 12. But each of us, as individuals, can play a part in bringing about these changes, doing our best to get as much food as possible from local markets rather than distant producers. Ultimately, what we are looking toward is supporting and building more sources of regional foods, while only eating foods from distant places to supplement local staples. Countries in the Northern Hemisphere, for example, cannot grow items such as coffee, tea, cocoa for chocolate, numerous spices, and so on. When we buy these imported products, we should choose those that are grown ethically and sustainably—fair-trade (meaning fair wages for foreign growers) and organic—so that our purchases are not contributing to the exploitation of another country's workers or natural resources.

Talk with Your Local Restaurateur and Grocer

If your favorite restaurant does not provide any local, sustainable options, or only a few, tell them you would like to see more. Chefs and owners usually appreciate customer feedback. Even fast food restaurants have been known to change their menu based on customer pressure—for instance, imagine the mountains that consumer pressure had to move to persuade McDonald's to offer a greater variety of salads and low-calorie options. Fortunately, more and more grocery stores are starting to recognize that "locally grown" is the new label with appeal. Some grocery store chains, such as Whole Foods Market, are making a concerted effort to sell the coveted local foods. Many even feature pictures and write-ups about local farmers beside the produce they supply.

One big New York grocer, Long Island's King Kullen, committed to buying Long Island's seasonal fruits and vegetables for its fifty stores. In 1999, it spent $100,000 on produce from Long Island farmers. In 2004, it spent $4 million. New Seasons Market, a grocery chain of six stores in Portland, Oregon, has a looser definition of local, using a "Pacific Village" label to denote foods from Northern California, Oregon, Washington, or British Columbia. Even in the wintertime, about half of its produce is from that regional "village."

If you have a store nearby that does sell local, sustainable foods, be sure to thank the store manager for carrying foods you can support. And if you'd like your local grocer to carry more local foods, be sure to ask the manager to sell locally raised meat and vegetables from independent family farmers, and request that the food be clearly labeled. For anyone who might be uncomfortable approaching a store manager, the Sustainable Table Web site (visit www.sustainabletable.org) offers a printable version of an "I Care" card, which lists the reasons stores should supply more local foods. All you have to do is sign the card and leave it with the manager. Because profit margins are so small, grocery stores will listen even if just a handful of customers ask for a certain product. But, as I said, if you convince the store manager to start selling a certain item, make sure you purchase the product.

Eat Seasonally

We have trained ourselves to plan meals around any food from anywhere in the world at any time of the year. Eating locally means that we need also to become reconnected with the seasons, to think about meals in the way our ancestors did, organizing them around the fresh market vegetables, around seasonal delicacies. An easy way to get started is to eat one local, seasonal meal a week. Make a social occasion of it, invite family and friends to help, organize a seasonal foods potluck where recipes and resources are exchanged.

One way to better endure the lean months of winter and early spring is to preserve the local harvest by freezing fresh fruits and vegetables and leftovers such as organic soups and veggie dishes. Then they can be eaten during the late fall or winter. (It's generally recommended that we eat our frozen leftovers and produce within six months.) Those with the time or inclination can bottle or can foods of many sorts so that there will be a huge variety during the colder months. Some enthusiasts have even started local food clubs where consultants, such as chefs who specialize in local cuisine or experts in kitchen gardens, home canning, and preserving, are hired to teach workshops.

Protect Endangered Foods

Those who are active in the Slow Food and local foods movements agree that it's important to support the farmers and artisans who are protecting endangered foods. Therefore, it is a good idea to include on our shopping lists those foods from other nations or states that are in danger of going extinct if the producers can't establish a larger market. For instance, the Tohono O'odham tribe plans to make its indigenous food products available for mail order. The Slow Food Web site offers numerous links to retailers who provide endangered foods. We also list international groups working on mail order food preservation in the Resources section.

Fresh, Handpicked, Homegrown

Of course the most local food you can eat is the food that comes from your own garden. During World War II, many U.S. and U.K. citizens recognized the need for creating more self-reliant food sources. People in all areas of the country, rural and urban, made Victory Gardens to help raise food for their families, friends, and neighbors. Nowadays, there's a resurgence of interest in homegrown foods, as more and more people are concerned about their health, and the vulnerabilities of a more globalized food supply.

If you live in an urban area and don't have land to plant on, you can often find community gardens, sometimes called "pea patches," that offer plots in exchange for nominal fees or volunteer time. More than 10 million urban dwellers raise vegetables this way in small gardens across America. Thirty-eight U.S. cities host community garden projects and one third of the nation's 6,000 community gardens were formed in the past decade. Similar systems have been established in Europe and at least some parts of Africa. (The Resources section directs readers to the American Community Gardening Association for information on finding a local garden.)

Lettuces are a great place to begin; they're easy and immediately satisfying. Tomatoes are another great starter plant because they are so much tastier when grown in your garden. Though it's tempting to want to begin with delicious heirloom tomatoes, more standard varieties are often easier to grow. Cherry tomatoes are especially vigorous, easy, and therefore satisfying. Start small-scale, focusing more on the preparation of good organic soil and just a few plants.

Learn Where Grocery Store Food Comes From

It can be fascinating to learn more about where our food comes from. How far did it travel to get to our plate? How was it grown, raised, caught, or killed? Studying the labels on the cartons or packages at the grocer should help us discover those products that come from a local producer and those that have been transported from far away. Perhaps our vegetables come from a local farm. What do we know about the farm? Does it use a lot of fertilizers and pesticides? If so, do we really want to eat those vegetables? Do we want to buy them for our children, our guests, ourselves? Perhaps the fruit we chose came from a foreign country. Do we know where that is on the map? What sort of lives do the people there live? These are especially important lessons for children: One of the projects for primary school children of the Jane Goodall Institute's Roots & Shoots global program for youth consists of students going through the ingredients of meals they ate at home, then discussing this in their group at school. It is one of the best ways of learning geography, and they acquire a lot of other information also in the process.

We need to build a closer connection with the food we eat because, after all, it does become integrated into our bodies, into our muscles and nerves and blood. We are made up, physically, of what we eat and drink. We must begin to choose our food accordingly.

Local Know-how

After the 2004 earthquake in the Indian Ocean created the terrifying giant tsunami, the Bay of Bengal's Andaman Islands were especially hard-hit. But because globalization hadn't wiped out their local foods or local knowledge, the Jarawa, Onge, and Sentinelese tribes managed to survive the giant wave. *National Geographic* writer Bernice Notenboom, who specializes in indigenous cultures, tells us that the tribe's awareness of the ocean, earth, and the movements of animals had accumulated over 60,000 years of inhabiting the islands. Oral history teachings helped prepare them to take proper action after they felt the first trembles of the earthquake. Their hunter–gatherer lifestyle, which depended on the wild food of the islands, allowed them to survive deep in the forest when the shore was hit by tidal waves. Moreover, no prawn farms—or tourist hotels—had destroyed the mangrove barriers.

In contrast, the Nicobarese people on the nearby island of Car Nicobar didn't fare as well, Notenboon says. Having assimilated the food and lifestyles of mainland India, the Nicobarese had lost connection with ancient teachings and allowed their forests to be cut down and turned into coconut and yam plantations. Being so close to the epicenter, and with no trees to protect them from the tidal waves, twelve villages were wiped out and many people were killed. Those who did manage to survive had a harder time coping in the aftermath because they no longer had the indigenous knowledge for living off the land.

Chapter 14 An Organic Wave Worldwide

"Humans merely share the earth. We can only protect the land, not own it."

—Chief Seattle

Why should we be concerned about what is happening around the world? Don't we have enough problems at home without worrying about Africa or India or China? Unfortunately, as corporate globalization stretches its grasp ever further across the face of the earth, we find that the demands of the urban elite seriously and negatively impact the poor in developing countries. As when the Brazilian rain forest is destroyed in order that cheap hamburgers can be sold on the American market. Or when large areas of traditional farmland in Africa are sold to foreign companies to grow coffee or tea, the sale of which is unlikely to benefit those who live in the area.

Looked at from another perspective, the government subsidies for farmers growing corn in America, which result in flooding the U.S. market and lowering prices to the extent that small-scale farmers are going out of business, also means that cheap grain is available for famine relief overseas. And while this may be of the utmost importance—in many cases saving thousands of lives—there can also be a downside in that local farmers in famine-stricken areas will be unable to sell *their* crops. Which may be devastating.

Finally, the amount of food shipped between countries has quadrupled since the 1960s, but this hasn't been a boon for the small local farmers in either the exporting or importing countries. Instead, it's perpetuated a situation where people in wealthier nations are consuming *more and ever more* of the food resources of poorer nations. We now have a global corporate structure where less developed nations are struggling with overpopulation, poverty, and hunger while they deplete their land and natural resources to feed people in other, wealthier parts of the world, and to put foreign exchange into the often corrupt pockets of government officials. And the small family farmers in the importing countries cannot compete with cheaper imported produce. Moreover, while the children in poorer parts of the world are often starving, even dying of hunger, the children of the developed world face an epidemic of obesity. Clearly something must be done.

SUSTAINING HOPE

All around the world we find hopeful examples of successful sustainable farming operations in poor, rural communities. And when we say sustainable, we mean, of course, the deep organic farming we've discussed, where the soil is nurtured through composting, biological pest control, and rotation of crops and livestock without the use of chemical pesticides or fertilizers. One such project linked farmers from the impoverished Makuyu community in Kenya with the Kenya Institute of Organic Farming. Before this partnership was established, the Makuyu farmers were using agrochemicals on depleted soil, and were struggling to grow enough food to feed their families. After learning sustainable, organic methods—similar to their old, traditional way of farming—they found that vegetable crop yields not only increased by 60 percent, they actually had a surplus of food.

But the good news didn't end there. The farmers decided to start a local food co-op, so they could sell surplus food and put the profits back into the community. As a result, the Makuyu co-op was able to buy dairy goats, beehives, rabbits, and poultry for community members as well as plant 20,000 trees, including 2,000 mangoes, to help revitalize areas that had been deforested. Meanwhile, the mood of the community shifted from despair to optimism. The Makuyu organic farmers have since gone on to teach other farmers in the region how to farm sustainably.

This is not an isolated example. Jules Pretty, director of the Centre for Environment and Society at the University of Essex and author of *The Living Land: Agriculture, Food and Community Regeneration in Rural Europe*, researched communities all over the world where farmers are replacing synthetic farm chemicals with sustainable, organic methods. He found that those who are no longer dependent on expensive imported farm chemicals can increase their yields while lowering their production costs. And because sustainable farming is often labor-intensive, it provides more employment for local and regional communities.

Pretty writes about projects that affect some 45,000 farmers in Guatemala and Honduras who now use organic farming and have tripled their corn yields. By also diversifying their upland farms, they created more local business and wealth, which helped spur a remigration back from the cities. He also reports that a million wetland rice farmers in Bangladesh, China, India, Indonesia, Malaysia, Philippines, Sri Lanka, Thailand, and Vietnam changed their farming methods to nonchemical sustainable agriculture, and increased their yields by about 10 percent.

But there is danger here. Increased food yield must be balanced by an optimization of the population in a given area. However carefully farmed, no land can produce enough food, by any method whatsoever, to keep pace with human population growth as it exists today in many parts of the world. When the number of people living in a given area is too great for the carrying capacity of that area they will try to move to new places. In many cases this is already impossible—there are simply too many people. If they are wealthy and can buy food from elsewhere, then they are depleting the natural resources in other areas. If we do not impose limits on our population growth, life as we know it, on this planet, will no longer be possible. Even if we could, theoretically, feed many more times the number of people than those on the planet today, how many of us would like to live on a planet where villages, towns, and cities meet and merge in one great urban sprawl across the face of the globe?

HELPING WOMEN TO TAKE CARE OF THE EARTH

In Tanzania, the Jane Goodall Institute has originated the TACARE (Ta(ke) Care) project in thirty-three villages around Gombe National Park. This has hugely improved the lives of more than 150,000 people by introducing fuel-efficient stoves for cooking, tree nurseries, methods of farming most suitable for the very steep rocky slopes bordering Lake Tanganyika, and ways to prevent or deal with soil erosion. All of our methods are, of course, based on organic, sustainable land use.

TACARE has established nine small microcredit banks (based on the Grameen Bank model) so that now small groups of women can start their own environmentally sustainable projects. Bright schoolgirls can apply for scholarships to enable them to go on to secondary school. And TACARE also offers women's reproductive health counseling, including family planning information and HIV-AIDS education. There is an emphasis on educating girls and women partly because, traditionally, their lives have been unacceptably harsh, but especially because it has been shown, all around the world, that as women's education increases, family size drops.

All the TACARE villages can now collect their supply of firewood from their own woodlots where fast growing species have been planted close by. And when they stop hacking at the tree stumps growing on the bare mountain slopes a new tree springs from the seemingly dead wood—within five years it will be twenty to thirty feet high. Now "TACARE forests" have sprung up around many of the villages. We are now taking the first steps to replicate TACARE in other parts of Africa.

IT ALL BEGINS AND ENDS WITH SOIL

More than 90 percent of the world's food comes from the soil. If you take into consideration that the food animals are sustained by plants, then everything we eat originates in the soil. Thus it is disturbing to learn from a recent U.N. report that each year more than 10 million hectares (25 million acres) of topsoil are swept away from cropland by rain and wind. Three hundred million hectares, an area that could potentially produce enough to feed the whole of Europe, have become so degraded that they cannot be used for agriculture, at least in the foreseeable future. "Farming," as Dr. Ward Chesworth of the University of Guelph said, "has produced an agricultural scar on the planet affecting one third of suitable soils."

This degradation is largely due to the clearing of woodland and forest for growing crops and gathering firewood to feed mushrooming human populations in the developing countries. In Ivory Coast, West Africa, for example, the loss of topsoil before clearing the trees was about .03 tons per hectare per year; after deforestation it was some ninety tons per hectare per year. India is losing some six billion tons of topsoil per year, again mostly due to deforestation. China, about the same size as the United States, and with three times as many people, has only one eighth as much good farmland. And this precious land, in many places, is turning into desert ever more quickly. The process has been going on for hundreds of years, but has intensified in the past fifty years, as population growth has led to the attempted cultivation of increasingly marginal land. There the thin soil soon dries up and blows away. The amount of land available for food production per capita dropped by half between 1950 and 1990, and since then, despite major efforts, the problem has only intensified. It often results in major dust storms: There were twenty-three in the 1990s. And in 2001 a dust cloud blew off China so massive that it briefly darkened the sky over North America.

And now China's farming problems are compounded as new urban and industrial developments radiate out from the cities. Farmers are losing their land, so that the percentage of the country suitable for agriculture is continually decreasing. In view of this, it is clearly desperately important to regenerate as much despoiled farmland as possible, as soon as possible. It is no longer appropriate for people, however rich or poor, to continue to destroy the planet's future. Yet it is easy to understand how this happens. In many places population growth has led to more people living in an area than the land can support, and as they struggle to scratch a livelihood they cut down more and ever more trees, often in places unsuitable for cultivation. This was the situation in the once forested hills outside Gombe National Park. By the early 1980s virtually all the trees outside the park were gone and cultivated fields stretched up toward the hilltops and to the boundaries of the park. During the wet season each heavy rain washed precious topsoil into the valleys and often directly into Lake Tanganyika. Once the trees from the higher slopes had gone, there were often flash floods. I visited one tiny lakeshore village where half the houses had been washed away and fifteen people killed during such a flash flood. The people around Gombe, like so many across Africa, were too poor to buy food from elsewhere. Some moved away, leaving family and friends to try their luck in the less crowded areas to the south. The others continued, desperately, to try to coax food from the increasingly barren land.

One of TACARE's most exciting projects helps people repair overused land, and to reclaim farms that were abandoned after deforestation, overuse, and erosion left them seemingly dead. Two demonstration farms, now lush, green, and with many trees, serve as models, and are popular with local farmers, who arrive in large numbers to learn new techniques. This fantastic ability to regenerate quickly is typical of many kinds of trees in the tropics, and there are methods of building up soil even in the most arid places, provided there is some rain.

As we have seen, the corporate-dominated, global food markets have a tendency to create plantations and industrial farms that degrade and pollute precious resources, such as the topsoil, water, and forests. Community projects such as TACARE can create a vast landscape of worldwide farmers who won't be bought out or compromised by the dangling carrot of immediate cash. Instead, the main incentive will be creating sustainable farms that best serve the farmers and their families, the land, and the consumers who are nourished from them.

LOCAL FOOD FOR LOCAL PEOPLE

Projects like these give us great hope. They restore degraded soil and safely increase crop yields all over the world. But one of the most inspiring outcomes of the growing interest in global, organic agriculture is that it emphasizes the importance of local food economies. Some skeptics might dismiss the local foods movement as a bourgeois food trend, serving those who have the resources and leisure time to eat delicious organic meals. But eating foods from local, sustainable sources is far more than a luxurious choice. It is our global mandate. Currently, 38 percent of the earth's land area is cropland or pasture and that amount is only increasing as the human population continues to grow. Some predict our supply of food will need to at least double and possibly triple over the next several decades to accommodate the planet's population growth. The use of toxic synthetic chemical fertilizers, pesticides, and herbicides, growth hormones, antibiotics in animal feed, food irradiation, and genetically modified organisms to increase food production has been justified, in part, by the rationale that without these products the world will not be able to feed itself. This is not the case—and even if it was, would a plentiful supply of contaminated food be the solution?

WHAT YOU CAN DO

When we buy local sustainable foods we support a new food paradigm where local communities reap the benefits of trade, rather than a few multinational corporations. This doesn't mean we have to stop all food trade or no longer buy special foods from other regions of the world. But it does mean shifting our priorities so that we create a staple of local foods wherever possible. This way, food is only imported in places where local resources can't sustain the community.

Buy Fair-Trade and Organic Imports

It is important, when we buy food from other countries, particularly from the developing world, that we make sure the product was grown and harvested in an environmentally and socially ethical way—which means buying fair trade and organic as much as possible. When any of us eats from local, organic food sources, we are less likely to contribute to the exploitation of another country's people or precious natural resources. Not every region or community can or should produce all its food. But it doesn't make sense for poor, underdeveloped areas of the world to grow cash crops for other countries when their own people go hungry. Nor does it make sense for wealthy countries to import cash crops when they are already growing the same kinds of foods in abundance.

Drink Ethical Coffee

If you are a daily coffee drinker, like I am, then every morning you can do something to reduce your chemical exposure, support safe farming practices, and protect our tropical forests. For instance, by purchasing shade-grown coffee, you invest in a crop that was grown by a farmer under the umbrella of a rain forest—protecting the jungles and even migratory birds of the world. If it's not shade-grown, your coffee could very well be grown on a clear-cut, factory-style plantation, completely dependent on agrochemicals.

Shade-grown coffee requires less and sometimes no chemical fertilizers because the plants that are part of the forest's complex ecosystem naturally add nutrients to the soil. Even water irrigation isn't necessary because the canopy of trees provides enough shade to slow the water loss from evaporation. In Peru, a farmer who grows coffee in the shade of a forest can earn 30 percent of his income from other sales, such as firewood, fruits, and medicinal plants—all natural features of the shade system. If you want complete assurance no agrochemicals were used on the crops, then certified organic is the way to go.

Fair-trade coffee means that you invested in a system that pays growers a "fair price" for the coffee painstakingly grown. Shockingly, companies that don't practice fair trade pay the average coffee grower less than $3 a day. Imagine—each day farmers have to feed their families, educate their children, and maintain their homes and business with the same amount of money that the average Americans pays for a latte. If everyone were to insist on only drinking ethical coffee, we would help rural farmers retain their dignity as well as ownership of their land (keeping it out of corporate control). We would reduce the amount of chemicals polluting the planet. We'd protect the future of migratory birds. And we could preserve almost 25 million acres of rain forest a year.

One cup at a time we can help reverse the trend of forest destruction. Forests and woodlands can and will grow back in most cases—not immediately and not exactly as they were before—but nature is resilient and eternally creative. So each time you buy a bag of beans or sip a cup of coffee you can fully enjoy its roasted, aromatic pleasures, knowing that you are protecting the world's tropical farmers and their precious landscapes.

Chapter *15* At Home and at School: Feeding Our Children

"The day is coming when a single carrot, freshly observed, will set off a revolution."

—PAUL CÉZANNE (THE EDIBLE SCHOOLYARD MOTTO)

There was a time, when my son Grub was small, that I used to love cooking. Especially at Gombe over a wood fire. We could not take food outside during the day because of the baboons. But after dark, when they were safely sleeping in the trees, I would light a small fire and make supper. Grub loved pancakes (crepes), especially tossing them. It was a real challenge to toss pancakes out there on the sandy beach—for they were inedible if they missed the pan! But Derek, my second husband, was very good at it.

One evening we were eating pancakes when Grub suddenly whispered, "Look at Crescent!" Crescent was a genet, a very tame one, who often came, along with civets and a variety of mongooses, to eat the scraps we left out for them each night. On this occasion she had approached from behind and taken one of the pancakes we'd cooked for seconds. As Derek and I looked round, there she was, moving away, head held high, with the pancake, which she had managed to fold in half, held firmly in her mouth. It was a marvelous sight, and reminded me of the time when my childhood dog Rusty had been spied running across the lawn with an entire chocolate cake in his jaws. It was funny—but shocking, too. It was right at the end of the war and my grandmother Danny had saved up precious, still rationed ingredients to make that cake for Judy's and my birthday (we were born on the same day, four years apart). So I raced into the garden, yelling at Rusty, and he dropped it. How many people, I wonder, have iced birthday cakes with a large bite taken out—and by a dog!

In her introduction to the 1953 version of the classic cookbook *The Joy of Cooking,* author Irma S. Rombauer advised women to stay calm "even if your hair straggles and you drip unattractively." Appearances are important, she noted, but not nearly as important as creating an aura of tranquility and grace. "A meal represents effort and money," she explains. "It is worthy of a dignified hour."

With all due respect, I would suggest that companionship, conversation, and fun should be the main goals of a shared meal. I look back on all the meals of my childhood with nostalgia. Cooked by my mother, Vanne, or my grandmother Danny in the small cramped scullery and eaten around the table in the kitchen, or, for special occasions, the dining room. Danny was a wonderful cook who refused to look at cookbooks, preferring to use her own feeling for amounts and consistency. Our meals were never fancy—we couldn't afford anything other than the basics. But they were always tasty—dishes like shepherd's pie, macaroni and cheese, spaghetti—and on Sundays usually a very small joint of roast beef with roast potatoes and carrots or peas. The meat came from cows that grazed the fields and the vegetables were grown without chemicals (no intensive farms in those days). We were expected to be punctual—clearly it was impolite to the cook if we let the food get cold. And we were hungry, I suppose. At one time there was a gong—it hung on a hook in the hall—and Judy and I loved hitting it with the little stick with the padded head. It was left over from the days when Danny ran a convalescent home and was used to summon the residents from their rooms.

At mealtimes we were supposed to behave properly, sit up straight, not eat with our mouths open, and drink soup and eat peas correctly. But meals at the Birches were almost always fun. We all talked a lot ("Don't talk with your mouth full" was a constant plea, forgotten or ignored daily), laughed a lot, teased each other—and did *not* sneak scraps under the table for Rusty! When Uncle Eric was home for the weekend Vanne made an extra effort to have Judy and me behave well. He was quite Victorian and would have absolutely agreed with the *Joy of Cooking* author Irma Rombauer's belief in the importance of "dignity and tranquility," which he certainly never got when he was with us. When we wanted to leave the table we would ask "Please, can I get down?" But Uncle Eric, instead, would ask "Please may I be excused?" which set Judy and me into fits of schoolgirl giggles because at school that was how we had to ask permission to go to the loo!

There have been staggering changes in society in many parts of the world since I was a child. So many children are raised in households where all the adult family members work—from choice or from economic necessity—and no longer have the time or inclination to spend the hours in the kitchen preparing the kind of meals I enjoyed as a child. Not only are children eating more and more junk food, with often disastrous effects on their health, but the gathering of families around the dining table is becoming increasingly rare. It is contributing to the breakdown of the family, which is one of the great tragedies of our times.

In almost every culture of the world, the family dinner has been a place to strengthen bonds through telling stories from the day and sharing ideas, while enjoying good food. These days, less than 50 percent of American and U.K.

families actually sit down for meals together on a daily basis, according to national studies. By the time children reach middle school many families have completely given up on eating together. This is bad news for our children, especially since solitary eating so often leads to a diet of sugar and snacks. Television viewing also hinders family nutrition. A Tufts University study found that families who regularly watch television during meals eat less fruit, fewer vegetables, and more pizza, junk food, and soda than those who turn off the tube during dinner. By contrast, families who make it a priority to slow down, unplug from all the electronic stimuli, and join together for a homemade meal are twice as likely to have five servings of fruits and vegetables a day as those who do not, and are far less likely to eat fatty fried foods and sodas, according to a Harvard Medical School study.

Family meals do more than boost nutrition. Research shows that children who frequently eat meals with their families tend to do better in school and have fewer behavior problems. They also offer some insurance during the risky period of adolescence. According to a University of Minnesota survey, teenagers who regularly eat meals with their families get better grades and say they are happier with their present life and are more optimistic about their prospects for the future. They are also less likely to smoke cigarettes, have substance abuse problems, become depressed or suicidal, or develop eating disorders.

The twenty-first century was supposed to bring so much opportunity and convenience to industrialized nations, but instead it's created a breakdown in dietary habits. Family relationships are suffering, our food is less nutritious, our bodies are becoming fatter, and everyone is hurrying about—working more and enjoying life less. In Japan, many adults are lamenting the "hotel family" phenomenon in which family members live in the same house, but seldom spend time or eat together.

While we attempt to dismantle the fabric of industrial agriculture perhaps we should, at the same time, question the lifestyle that puts speed and convenience above well-cooked and nutritious food, eaten together as families. Shopping in the supermarket is quicker than visiting individual stores (many of which have gone). Package labels inform us that the various meals that can be heated up in the microwave are also healthy—when often they are not healthy at all. They are bought because it will save the bother of cooking, and because it saves time. This increase in speed, convenience, and calories has not brought us closer to our children, or increased the stability of marriages—and it certainly has not increased the quality of our health and nutrition.

As children are so often deprived of the chance of socializing around the family dining table, and the quality of the food they eat at home becomes increasingly bad for their health, it would be consoling if we could be assured that their nutritional needs were being met at school. Alas, this is very seldom the case.

SCHOOL MEALS

I remember school lunches very well—simple, plenty of it (you could have seconds), and, while not terribly well cooked—watery cabbage and lumps in the gravy—nutritious. Some kind of meat (not much) most days, or a cheese dish (not very cheesy), and fish on Fridays. I remember pressing down with my fork on a too long section of rhubarb—and a long caterpillar, dyed the pink of the rhubarb—popping out one end. Of course we all shrieked and made sounds of horror, as schoolgirls do. But that had nothing on the day when I sat at the high table—we did this in turn, under the eye of the headmistress—and noticed a very small slug on a leaf of her salad. Nudging my neighbor

I pointed this out, and we watched, in fascinated horror, as, eventually, she put the affected leaf into her mouth and chewed it up. I felt too shocked to laugh, and I can't think why we didn't dare tell her.

When we talk about school lunch it's tempting to begin with the scary stuff: Low-grade mystery meats. Side orders of gray-green canned peas. Government surplus milk from cows reared on Bovine Growth Hormones. Desserts of Jell-O squares. And that was the good old days.

Nowadays, school lunch is likely to be burgers from McDonald's or pepperoni pizza from Domino's washed down with a super-size Coke. As physical education and sports programs are being cut to meet budget constraints, school districts are handing over lunch contracts to fast food chains such as McDonald's, Domino's, or Taco Bell. Many schools are also signing "pouring" contracts with soft drink corporations, giving them the ability to both sell and promote their products on campus. These contracts offer big money for desperate school districts, with exclusive contracts from Coke or Pepsi often padding district budgets by millions of dollars. Almost two of every three middle schools and high schools nationwide sell soft drinks, mostly through vending machines. Worst of all, more and more children are becoming obese at younger and younger ages. We cannot absolve the parents, but we cannot absolve the schools either. And the worst offender, unquestionably, is the fast food industry. We shall discuss this in Chapter 16.

But when it comes to talking about school lunch, the truly alarming discussion is the rampant childhood obesity. Over the past thirty years, the rate of obesity in the United States has more than doubled for preschoolers and adolescents, and it has more than tripled for children ages six to eleven. Obese children are not only teased and have a hard time participating in physical activities; they are at risk for major health problems such as diabetes and heart disease, often carrying these risks into an obese adulthood. The American Heart Association issued a strong warning in 2005: Childhood obesity is such a critical public health problem that it threatens to reverse the last fifty years of progress against heart disease. If this trend continues, obesity will soon top smoking as the nation's most preventable cause of death.

When doctors and activists look to the root of the obesity epidemic they usually point to the effects of a fast food diet that emphasizes super-sized portions of high-calorie, poor nutrition foods filled with saturated fats and sugar. It is sad and disturbing enough that so many households no longer prepare healthy home-cooked meals from fresh ingredients, but it is criminal to think that our public schools are selling out our children's health, however desperate their attempts to make ends meet.

Ironically, the first National School Lunch Program in the U.S. started after World War II to improve the nutritional status of America's children as future soldiers. Nowadays, the school lunch program is part of a downward trend in nutritional standards that is not only seriously compromising the health of our children but encouraging poor food choices that will probably last a (shortened) lifetime. Research shows that poor childhood nutrition is also linked to developmental problems, such as learning disabilities, aggression, and antisocial behavior.

A DELICIOUS REVOLUTION

Imagine how much healthier our children might be if we held healthy eating habits to be as important as math skills or history knowledge. Imagine if we taught children how to grow nutritious, organic fruits, vegetables, and herbs, and how to turn these foods into delicious, healthy meals. Imagine if children learned a happier, more wholesome story behind food, how it can be grown in service to the health of the earth, the body, and the human spirit. Picture a world where school curriculums taught the value of food traditions and rituals, the art of setting an inviting table, and the importance of mealtime conversation.

Sound like a dream? Well it was, and it belonged to Alice Waters, the iconic proprietor of Berkeley, California's, Chez Panisse restaurant. Having already based her award-winning restaurant cuisine on fresh, seasonal local foods from sustainable sources, Waters had a vision of bringing her delicious, healthy world of food to the public schools. So it seems only fitting that Waters's dream for America's children eventually took seed at a middle school named after Martin Luther King, Jr., just a few blocks from Chez Panisse.

After cultivating the support of the Berkeley Unified School District, Waters and some friends acquired a parking lot adjacent to the school's playground. Joined by over a hundred volunteers, she turned over an entire acre of asphalt, transforming the soil beneath it into a fertile garden that cultivates the antithesis of burgers, Coke, and fries: arugula, asparagus, kiwi, artichokes, red Russian kale, grapes, pumpkins, herbs, flowers, and much more. In keeping with the school's cultural diversity, each crop is identified with a sign written in the language of the student who was chosen to make the sign—which could be any of the nineteen languages spoken at the school.

Waters also helped fund the creation of a spacious, colorful, and fully equipped working kitchen classroom, built largely out of renewable materials. She also gathered up a small but highly competent staff to further flesh out the innovative program of study and, deftly blending all these ingredients together, created the Edible Schoolyard. Just as Chez Panisse became an inspiration for a national trend in locally sourced sustainable cuisine, this dream curriculum has since become one of the most influential and inspiring food stories of our time.

Among the more inventive elements of this groundbreaking project is that for the first time students were *required* to eat a nutritious lunch—and they received academic credit for doing so. Of course, they are also expected to take part in tilling, feeding, and weeding the garden; harvesting the crops and gathering eggs from the henhouse; cleaning and preparing the food in the kitchen; and cooking freshly prepared meals for themselves and their fellow students—all the while learning about sustainable ecosystems and resurrecting the civilized tradition of actually engaging others at one's table in conversation.

In a fast food world, where children believe that salads come from plastic bags, macaroni and cheese comes from a box, and fruits come in the form of colorful boxed candies, just spending time in a garden can be an eye-opening experience. "Most kids are disconnected from the sensory world of food," says Waters, "and our whole consumer society puts a barrier between children and the touch, smell, and feel of real life."

One telling example of this disconnection between inner-city children and the roots of their food is the comment made by a resident of Chicago's Interfaith House for the homeless. Upon learning about vegetables, this resident turned to a staff member and said, "You mean, all this time I have been hungry and sometimes have had to go without food, and now I find out food grows in the ground?" This might be excused in the case of this victim of the concrete jungle. But recently, when I was talking to six eight- to ten-year-old children in rural Nebraska, I was horrified to discover that only one of them knew that potatoes were dug up from the ground. None of them knew how artichokes grew, and most of them didn't know about cucumbers, peppers, or squash. Only one could name a kiwi and only one knew the meaning of organic food. Soon after this, at a checkout in a supermarket in Washington, D.C., a young assistant held up a grapefruit and asked his supervisor what it was! And when, in an attempt to improve school lunches, children in the U.K. were given whole apples or oranges, they did not know what they were—they had never handled a whole fruit in their lives.

I feel desperately sorry for these children who have no knowledge of the earth and its bounty. There is something incredibly special about food that you pick or collect yourself. My sister grows all kinds of vegetables in our garden in Bournemouth. There was a variety of golden cherry tomatoes that produced a never-ending crop of sweet fruits last year and eating them, warmed by the sun, was sheer joy. The taste of young peas, straight from their crisp green pods, is almost unbeatable. No pesticides in our garden—we collect the snails and slugs and take them far away, where they can do no harm.

When I was a child the only strawberries I really liked were those picked and eaten in the open air—best of all the wild strawberries—miniature containers of the essence of strawberry. And as for blackberries . . . just writing the word brings vividly to mind the blackberrying excursions of our childhood. My grandmother Danny was an avid blackberry hunter. Armed with a crooked walking stick she would tackle the most daunting thickets, risking innumerable scratches, to reach the biggest and juiciest fruits—always just beyond the reach of us children. I have introduced my grandchildren as well as my sister's grandchildren to the joys of blackberry picking along the cliff tops near home. What is so surprising—and a pointer to the way we have lost touch with nature—is the fact that these days we have the crop almost to ourselves. People look at us as though we are crazy—eating fruit from wild plants. They seem not to understand. How sad, for our youth is missing so much.

Waters knew that the more inner-city schoolchildren saw the process of growing and preparing food, the more healthy and empowered they would become. "We teach children that carrots don't come from the supermarket, they come from the ground," says Waters. "And preferably even show them how they grow in the ground." But she also knew that all the lessons of growing organic fruits and vegetables would fall on deaf ears if she couldn't get children to enjoy them. That's why she made sure that the Edible Schoolyard also taught children how to cook the garden-fresh food in ways that were better than appetizing—they had to be irresistible. "I call it a Delicious Revolution," says Waters. "I'm not asking these schoolchildren to read some difficult philosophical book about ecology and nutrition. I'm asking them to go out and eat in the garden, and to pick something that's beautiful and tasty and eat it."

Like the abundant organic garden, the Edible Schoolyard's teaching kitchen is an antithesis to fast food. Instead of convenience, the kitchen emphasizes tradition and art. You won't find any Cuisinarts, electric can openers, or a microwave. The tools of choice—wooden spoons, mortars and pestles, old wooden tortilla presses, and various kitchen knives.

Here the children learn to cook these exotic fresh foods from the garden: golden beets, collards, tatsoi. They learn to fall in love with flavors and smells—the simple scent of sautéed onions, the harmony of savory sage in a soup filled with sweet carrots. In contrast to a week of Wendy's, Domino's, and Taco Bell, an Edible Schoolyard week's menu could include Jerusalem Artichoke Fritters, Pumpkin and Kale Soup, Cucumber Sushi, Sweet Potato Biscuits, and Brown Rice Salad in Red Chard Leaves, all washed down with iced tea made from the garden's verbena and hibiscus.

Once the food is prepared, the pleasure principle extends to the dining room. No Formica tables and fluorescent lights allowed. Instead, the tables are hand-milled from recycled California hardwoods with matching benches and stools. And tablecloths and nice silverware, properly laid. "They appreciate the beauty of the floral tablecloth," says Waters. "They notice how much pleasure fresh flowers bring to the table and how beautiful the room looks with the lanterns lit."

Many of these children don't participate in sit-down family dinners, so all the art and etiquette of sharing a meal becomes an extension of the classroom. As they sit down together to share the fruits of their labor, the children are given daily question cards to prompt interesting conversations while dining. Sometimes the lunch feasts are linked to historical or cultural studies in other classrooms. For instance, Mexican culture is brought to life with a "Bread of the Dead" lesson where students make the traditional bread used on Day of the Dead altars in Mexico. Students also create altars using flowers and herbs from the garden and write remembrances of loved ones who have died, posting them on a bulletin board. As it happens, nearly every student at Martin Luther King, Jr., Middle School has experienced a personal loss through death (many due to violence). Learning the food and rituals of the Day of the Dead offers students a safe and structured way to express and process these losses. And it also offers the Mexican-American students a chance to share their enthusiasm and expertise about their culture.

A Delicious Revolution may seem like a frivolous gourmet luxury. Yet teaching children how to prepare and enjoy these diverse foods of the earth is the foundation for stopping the obesity crisis in its tracks while also saving the planet from the ravages of industrial agriculture. This could be the revolution that eventually changes the face of global eating. If children fall in love with the fresh, organic foods of their homeland, if they learn their names and how to prepare them, if they learn the rich cultural traditions of foods, they will carry that knowledge and appreciation well into adulthood. Even in rural communities, market research shows that one of the main reasons fresh foods are not bought is not the cost or the availability, but because consumers do not know how to prepare them. Imagine if we created a whole generation who not only knew how to make homemade feasts from kale and golden beets, but fiercely protected their right to do so.

For now, Waters sees the Delicious Revolution as a healthy antidote to a fast food world. "We have fast food music, fast food art, fast food architecture," says Waters. "By that I mean it's all fast, cheap, and easy, and nobody's considering the environmental or cultural destruction. Kids are absorbing these values and it's alarming. We aren't paying attention because our fast food nation has taught us to consume and dispose."

But in the school's kitchen children learn a new relationship with the earth and food—eat sustainably, recycle more, and make your footprint on the earth as small as possible. Vegetable peels and scraps are matter-of-factly turned into kitchen stock or compost, a tin can is made into a cookie cutter, bottles are recycled into rolling pins. At every turn children are taught that the decisions we make about food affect the health of the planet. So far the Edible Schoolyard program hasn't seemed to cause a remarkable improvement in academic grades, but teachers frequently credit the improved nutrition with better behavior in the classroom, according to Marsha Guerrero, executive director of the Edible Schoolyard. Having students work side by side in the garden also creates a more amicable social scene. And it gives those who don't succeed as easily in the classroom opportunities to shine in the garden, kitchen, and dining room.

AND THEN THERE'S THE BIG DREAM

Waters still dreams that someday every school will have an Edible Schoolyard program and give academic credit for eating a healthy lunch. "At the very least I'd like us to raise children who know how to grow and cook simple, affordable food while living on the planet in a sustainable, compassionate way," she says. For now, she's taking it one step at a time. It may not be realistic to have an accredited garden-to-lunch program at every school, but it *is* realistic to start weaning schools off unhealthy, fast food meals and start hooking schools up with local, sustainable foods. So far her vision is becoming a reality in Berkeley, California, where the school district is considering a partnership with the region's numerous sustainable farmers so that 10,000 children can have deliciously prepared, farm-fresh food served to them every day. Next stop—the entire state of California. Considering that the bill for California's obesity problems is estimated to be $21.7 billion a year, it's not surprising that she has the backing of numerous fiscally conscious state leaders, including Governor Arnold Schwarzenegger and First Lady Maria Shriver.

Meeting Waters, it's hard to imagine that she couldn't single-handedly transform the world's public school meals into a source of nourishment, environmental awareness, and global health. Fortunately, though, she doesn't have to do it alone. All over the nation teachers and school officials are realizing that most children have lost their connection with food and the land. While the stereotype of Berkeley as a liberal enclave makes it seem a likely place for schools to offer local sustainable foods in their lunchrooms and curriculums, similar programs are sprouting up all over the U.S. and in many European countries as well.

School Lunch Around the World

A typical school lunch in *Japan* is simple but highly nutritious—a bottle of milk, a bowl of rice, a fish dish, pickled salad, soup with vegetables and tofu, and a piece of fruit. Teachers eat alongside their students, socializing while also discouraging picky eating and wastefulness.

In Helsinki, *Finland,* the school meal menu is posted four weeks in advance on the board of education Web site. Typical entrées include ham and potato casserole or barley porridge. Children always have a vegetarian option, such as coconut milk and beetroot casserole. One of the students' favorite meals is spinach pancakes.

In *Spain* children often bring home a list of their school meals in the beginning of the week, with every meal broken down into the amount of calories, as well as the fat, protein, carbohydrate, vitamin, and mineral content. School districts even offer suggestions for a complementary evening meal to ensure that children have a day of well-balanced nutrition.

Italian school meals are considered some of the best, with a national mandate that all schools have organic offerings on the menu. The schools began emphasizing a Mediterranean diet sixteen years ago, which means developing menus with less red meat and more fish, seasonal fruits, vegetables, and whole foods. Lunchtime should last forty-five minutes, and often includes fresh flowers gracing the tables.

Lunch in primary and secondary schools in *France* costs the local school districts anywhere from $3 to $7 per child—and no one seems to doubt whether this investment is worthwhile. School meals always include an appetizer (such as a grapefruit half), a main course of meat or fish with vegetables, a dairy product (slice of local artisan cheese), and dessert. Some schools have food groups color-coded—one color for dairy products, one color for fruits, another color for vegetables, and so on. Students are required to take a serving of each color. Anything less than a one-hour break for lunch is considered barbaric.

FEED ME BETTER: THE NAKED CHEF GOES TO SCHOOL

The U.K. may not have the elegant and articulate celebrity chef Alice Waters, but it does have the dynamic, flamboyant, and equally charismatic Jamie Oliver, the colorful chef for the *Naked Chef* television series. Oliver somehow got to know about the sorry situation with British school lunches (or "dinners" as they are called in the U.K.) and decided to use his celebrity status to do something about it. The British government funds a hot meals program for all children in public schools—which sounds great until you see what's served for dinner. A standard meal might be Turkey Twizzlers—corkscrew-shaped items made of mysterious fillers and preservatives alongside a pile of fatty French fries, altogether making Wendy's seem like haute cuisine. But at a cost of 37 pence (about 70 cents in U.S. currency) per child per meal (less than is spent by most other European Union countries), it's hard for the school caterers to do much better.

Much of Oliver's campaign—called Feed Me Better—is to get more government funding for schools while working with "dinner ladies"—school kitchen staff—to improve the quality of foods. Oliver's new television series, *Jamie's School Dinners*, features shocking coverage of the unrecognizable, greasy glop that's contributed to the nation's soaring obesity problem as well as schoolchildren developing all kinds of shocking ailments, such as chronic constipation and impacted colons. He also notes that teachers often report that behavior problems peak right after lunchtime as students try to digest a lunch of processed food that's high in sugar, salt, fat, and a cocktail of additives. Like Alice Waters, he also finds it sad that children are losing the social graces that come with civilized meals. According to Oliver, many British schoolchildren don't even know the proper mechanics of using a knife and fork.

But his message is also hopeful—showing television viewers the variety of fresh, healthy foods that can be served in schools on a tight budget. Not only is he hoping to get the government and schools to dish out healthier fare, Oliver also hopes to teach children the roots of their food.

In the first episode of his new series, Oliver shows primary school students a bunch of celery and asks them what it is. Not a single child knows. But they all recognized the logos of burger and pizza chains.

Oliver's campaign has had a huge national impact, with Prime Minister Tony Blair backing a new government promise to spend an additional £280 million to improve school meals. Parents all over England are also pressuring schools to offer more farm-fresh foods, many are even demanding that schools provide organic and vegan menus. In response, more and more schools are mandating that fresh-cooked vegetables and salads be served each day and chips (the English version of French fries) only be served once a week. Others are banning Turkey Twizzlers and mechanically recovered meats. And some have gone as far as to insist that all beef burgers be organic. Britain's biggest teachers union boss, Steve Sinnott, thinks that chef Jamie Oliver should be knighted. We absolutely agree.

APPETITE FOR A NEW CURRICULUM

Recognizing that nutritious meals could help reduce health and behavior problems, many school districts are eager to serve meals made from fresh, wholesome ingredients. When the Los Angeles Unified School District began offering a Farmers' Market Fruit and Salad Bar at fifty-five schools, the district discovered that caloric intake fell by 200 calories, and fat intake fell by 2 percent. That's exactly what officials were hoping, especially because many of the district's children are from Mexican-American and African-American families, who are at risk for obesity and nutrition-related problems, including diabetes.

The biggest hurdle is the cost. Organic food, produced without pesticides, growth hormones, or other additives, generally costs more. That's a tough sell when schools can barely afford the cost of books and teacher salaries. The good news is that schools are coming up with unique ways to make fresh foods more affordable. For instance, by eliminating dessert items, Lincoln Elementary School in Olympia, Washington, managed to cut the cost of school lunches by 2 cents a meal while also offering an all-organic menu—with much of the produce supplied by local farms. "We have our precious children as a captive audience," says Lincoln principal Cheryl Petra. "It's only ethical for us to feed them the best and most nourishing food possible." Petra reports that parents and even students have enthusiastically supported the program. "They get enough sugar—it's not as though they really needed more at lunch." The elementary school also recognizes the health advantages of eating less animal products, offering a nonmeat source of protein with every meal.

Other schools have reduced the cost of fresh foods by taking a Community Sponsored Agriculture approach to local farms—directly contracting with sustainable growers to provide fresh ingredients to the school kitchens. Of course a diet of farm-fresh foods means a healthier diet. But the new nationwide Farm to School program also supports the health of communities by creating a secure market for sustainable growers, thus preserving the vitality of local family farms. Many schools who partner with farms also offer a modified version of the Edible Schoolyard, using the partnerships with local growers to arrange farm visits, teach nutrition and cooking classes, and even get advice on starting school gardens.

It's not just parents and nutritionists who are insisting on healthier meals. College students are getting fed up with the poor quality of institutional food as well as the "Freshman 15"—the popular name for the initial weight gain experienced by many students when they start college. Adriane Dellorco, a recent graduate of Oberlin College in Ohio, started campaigning for local food options in the campus dining service during her first year. After spending most of her college career trying to broker a deal between Oberlin officials, dining service managers, and local growers, she finally got results. About 5 percent of the school's food budget now goes toward local farms and distributors, about a third of which goes toward organic products.

Not surprisingly, when Alice Waters's daughter Fanny began attending Yale University, she immediately became instrumental in getting a fresh meals and garden program started at the university's residential halls. In the wake of student pressure, colleges and universities all over the U.S. are following suit. Cornell University's Farm to School pilot project inspired several New York school districts to start serving local produce, including fresh apples, cabbage, onion, tomatoes, potatoes, cucumbers, green peppers, carrots, cauliflower, broccoli, pears, and milk.

Hurdles remain. For instance, since many schools began to outsource premade, packaged lunches to fast food franchises and other suppliers, many facilities don't have the kitchens or staff to accommodate food being made on the premises. But those in the forefront of the local foods movement remain committed. "Health, environmental, and cultural problems are so serious in this country that there's an opening here," says Alice Waters, who points to the 1960s when the nation realized it had to help children become more physically fit and began building gymnasiums and funding physical education classes.

WHAT YOU CAN DO

While an Edible Schoolyard may be beyond your district's current financial abilities, a Farm to School program could be an ideal start. In 2000, the USDA began supporting the Farm to School movement with a substantial grant. The 2002 Farm Bill directs school food service officials to buy locally whenever possible. If you'd like to initiate a Farm to School program in your school district or even work toward an Edible Schoolyard, we list Web sites and organizations in the Resources section.

A number of Roots & Shoots groups across the United States and Europe are growing organic gardens at the more affluent schools, learning about composting, and delivering the fresh produce to the elderly or the homeless.

Break the Unholy Alliance

Even if you can't get local farm food into your school, many parent groups have been able to influence school boards to break the unholy alliance with fast food chains and soft drink companies. Parents have been especially successful in exchanging soda machines with vending machines that offer juices, water, and dairy products. You can even ask schools to ban snack vending machines that typically offer low-nutrition food such as candy and potato chips and replace them with healthier vending machines. Stonyfield Farm (a company that specializes in hormone- and antibiotic-free dairy products) had the good sense to create a new kind of vending machine for schools that offers low-fat yogurt, string cheese, organic milk, carrots and dip, dried fruit, raisins, pita chips, and other healthy snacks. All items have to meet the nutritional standards established by the school and must be taste-approved by students.

The main point here is that parents can and must use their influence to protect children's health. You'll be surprised how many other parents as well as teachers and students will support you in getting healthier foods available in your local school.

Chapter 16 Obesity, Fast Food, and Waste

"But why should these companies want to change? Their loyalty isn't to you, it's to the stockholders. The bottom line: They're a business, no matter what they say. And by selling you unhealthy food, they make millions. And no company wants to stop doing that. If this ever-growing paradigm is going to shift, it's up to you."

—MORGAN SPURLOCK, *SUPER SIZE ME*

The problem of obesity is not new. The image we have of fat, jolly medieval monks is, in some cases, entirely accurate. In the thirteenth century monkish obesity was widespread in Europe. One Portuguese order even devised a test: Monks who were unable to squeeze through the doorway into their dining room had to fast until they could! Skeletal remains from monks who lived from 476 to 1450 show that most of them were significantly overweight and actually suffered from many ailments we associate with obesity, such as type II diabetes, arthritis, and back problems, according to a study by archaeologist Phillipa Patrick. Of course, obesity was unusual in medieval times, since all but the very wealthy were often undernourished. But many monasteries figured out ways to hoard food (some were accused of stealing the alms for the poor and using it for their gluttonous feasting habits). Typical monastic diets included plenty of fruits and nuts, and a few vegetables. But they also ate a lot of animal products, such as meats, milk, butter, eggs, and cheese. Because their diet was high in saturated fats and they lived such a sedentary lifestyle, it is not surprising that their illnesses mirror many of the health problems we now see in Western culture.

A Medieval Monk's Diet

The following is an example of a thirteenth-century friar's possible daily intake, partly based on a study of skeletal remains of medieval monks from Tower Hill, Bermondsey, and Merton abbeys:

11 A.M.–1 P.M. Three eggs, boiled or fried in lard. Vegetable porridge with beans, lentils, carrots, and other garden produce. Pork chops, bacon, or mutton. Chicken, duck, or goose with oranges. Half a pound of bread to use as sop. Peaches, strawberries, or bilberries with egg flan. Four pints of small (watery) beer.

4–6 P.M. Mutton gruel with garlic and onions, a milk shakelike "posset" of egg, milk and figs. Venison, with rowanberries, figs, sloes, hazelnuts, and apples. Stewed eels, herring, pike, dolphin, lampreys, salmon, cod, or trout. Half a pound of bread as sop, sometimes soaked in drippings or lard. Syllabubs of fruit. Four pints of ale. A flagon of sack or other French, Spanish, or Portuguese wine.

The ample proportions admired in the most respected and wealthy citizens in many African countries fall far short of obesity. There it is a sign of wealth, and to be encouraged in the wives of the wealthy to demonstrate the husband's bounty. However, explorers from the late 1800s discovered that the king of Buganda used to fatten up his wives, taking the most beautiful virgins and forcing them to consume vast amounts of milk and honey until they could no longer walk. They had to be rolled to him for his pleasure. The mind, as they say, boggles.

If we take a quick cruise around the animal kingdom, we find that the most well-covered animals are those that must endure extremes of cold—such as whales and sea lions who surround themselves with a protective layer of blubber. This certainly does not impede them in their element—water. Some animals, such as bears, fatten up for hibernation. But they are far from obese. Only domesticated animals, or wild animals in captive conditions, become obese.

We are all familiar with those sad, overweight dogs and cats, overindulged, overfed—and literally being killed by kindness. Only, of course, it is not kind. The point is, there is no mechanism that turns off the desire—instinct, really—to eat food when it is available. From an evolutionary perspective it makes perfect sense. In the wild, carnivores must hunt, and they may not always make a kill every day. I have often watched lions, hyenas, and jackals gorging on a kill until they can barely walk. This is the equivalent of a person who lives a long way from town stocking up on food for the next week or so. In the case of the wild creatures, the stomach acts as the refrigerator. And so our dogs and cats, acting out their ancient program for survival, eagerly gobble up any delicacies that come their way.

Until recently, people knew little of the natural diet of chimpanzees (and other wild animals), so we sometimes encounter overweight individuals in captivity. They cannot resist feasting at every opportunity on food they enjoy. And so it is in the wild—when a fresh crop of food ripens they sit and gorge themselves. They need to, for there is competition for seasonal delights, not only from other chimpanzees, but also from baboons, other monkeys, fruit-eating birds, and a host of small mammals. And, as with the carnivores, it is important to stock up for the lean times of food scarcity that can strike at any time. The difference is that in the wild they burn up a lot of fat in their daily lives. It is unlikely that a chimpanzee in the wild could become obese even if he or she so desired!

It seems equally unlikely that our prehistoric ancestors ever suffered from obesity, any more than the modern hunter-gatherer tribes. For the most part it is only in the affluent societies around the world, when people have too much of everything, or in urban societies where people are encouraged to eat fast food, that obesity is widespread. Recent research showed that whereas obesity, for many years, was associated with those in the lower income bracket, today it is a problem also for the wealthy. It is a condition that crosses all socioeconomic barriers.

Indeed, obesity is being described as an epidemic in America, the U.K., and parts of Europe and Asia, affecting at least 300 million people. In the U.K. more than 66 percent of adults are considered obese. In the U.S. 30 percent of adults—that is, more than 60 million people, are obese (one in three women and more than one in four men). Obesity in the U.S. is responsible for $100 billion in medical costs and some 300,000 deaths annually, according to the American Obesity Association. And the extra body weight of passengers meant that airlines burnt 350 million more gallons of fuel in 2000 than in 1980.

The obesity statistics for children are especially disturbing. In the U.K. 8.5 percent of six-year-old children are obese, as are 15 percent of fifteen-year-olds. In the U.S. childhood obesity is growing at the rate of 20 percent per year, with about 16 percent of the children and teenagers considered overweight.

This very serious epidemic is almost certainly linked to the increasing consumption of junk food and fast foods. Each day, between 20 and 25 percent of the American public eat in some kind of fast food restaurant. And it doesn't help that so much of the advertising of junk food and fast foods unethically targets children. In the U.S. on any given day, 30 percent of American children ages four to nineteen eat fast foods.

THE NOT-SO-MISSING LINK

People who start buying fast food or processed junk food because it is cheap and quick to prepare (sometimes you just need to open the package) soon come to rely on it—with disastrous results to their health. And unfortunately there are giant corporations that spend thousands of dollars on developing, packaging, and advertising fast foods and junk foods to the detriment of consumers but to the benefit of their shareholders.

The leading causes of death in the United States are chronic diseases associated with excessive (or unbalanced) intake of food and drink, according to Marion Nestle, chair of the Department of Nutrition, Food Studies, and Public Health at the Steinhardt School of Education at New York University, and author of a fantastic book called *Food Politics: How the Food Industry Influences Nutrition and Health.* She compares the fast food industry to the tobacco industry—both are highly profitable, have powerful allies in the government, and show a blatant disregard for the harm that their products do to consumers.

THE SEDUCTION OF HAPPY MEALS

Worst of all is the rampant childhood obesity that comes from feeding our children an unhealthy diet of high-fat, high-sugar, fast food garbage. Many people questioned the validity of a class-action lawsuit filed against McDonald's in 2002 by New York teenagers Jazlyn Bradley, then nineteen, and Ashley Pelman, then fourteen, who claimed that the food giant's food made them obese and suffer from heart disease, diabetes, and high blood pressure. Couldn't these children have just refused to eat so much fast food? But looking more closely at the lawsuit, we find remarkable similarities to the tobacco industry lawsuits. These teenagers claim that it was the unhealthy *content* of the food, not the amount of food eaten, that put them at risk for serious health problems. Although the teenagers' case was originally dismissed in February 2003, it was reinstated in January of 2005 and has yet to be determined.

Their case may have a greater chance of success because of the message of an extraordinary documentary film, *Super Size Me*, which charts the health of filmmaker Morgan Spurlock, who, for one month, lived on nothing but McDonald's food. He followed three simple rules: 1) He could only eat what was available over the counter; 2) He could only supersize when he was offered; 3) He had to eat every item on the menu at least once. Although he started off with a clean bill of health after medical checkups with three different doctors, by the end of the movie he had gained twenty-five pounds, was suffering from chronic headaches and nausea, his moods were swinging back and forth between lethargic depression and manic overdrive, and his liver and heart became so impaired that his doctor was begging him to give up the experiment even before the end.

Monica Ferreria, a Roots & Shoots coordinator who works with young people in Salt Lake City, recently spoke with me about one area that is often overlooked: the improper nutrition of the struggling poor in the United States. Yes, they do have food, she explained, but they are being poisoned by the quality of food. She told me about stores where everything is a dollar. These stores are very popular with young mothers who have small children and not very many resources. The food is cheap, but it has horrible nutritional value. And Monica said, "Everyone loves Twinkies and potato chips—the marketing of these products to children is unethical. It is an interesting dilemma, when the poor are not starving from lack of food, per se, but from *too much* of the *wrong food*."

Exactly so. Indeed, it becomes increasingly clear that the giant corporations behind the fast foods and junk foods, which are inundating the market around the globe, are motivated by making profit rather than providing nourishment. So it is up to us—the consumers—to bring their unethical business practices to an end.

With so many people suffering from diet-related illnesses, consumers are beginning to realize that the fast food industry should be held accountable for its global agenda to produce cheap, low-quality, high-calorie food for profit regardless of the harm it causes. As we saw in the last chapter, our children are especially at risk because of the quality of school meals, and the shocking capitulation to corporations such as McDonald's who are now providing the meals, and have demanded, as recompense for "good" deals, the right to install vending machines for their products. Indeed, there are schools that encourage their students to work for the good of the school by collecting soda caps. Schools with good scores are rewarded with substantial cash prizes. In addition, the fast food chains specifically target their advertising toward children who are seduced by Happy Meals and packages with free toys, games, collecting cards inside, and who eventually come to prefer the excessive sweetness.

BEWARE OF CORN SYRUP

Corn is the most prevalent crop grown in the United States—some 78 million acres of farmland have been planted in recent years. And no USDA subsidy program sends taxpayer money to more recipients. Between 1995 and 2003, just under 1.5 million individual farmers, partnerships, corporations, estates, and other entities received at least one corn subsidy payment. Michael Pollan, the author of *The Botany of Desire*, believes that the U.S. subsidizing of corn crops is one of its most harmful agricultural practices. It has caused enormous damage to the environment as industrial corn farmers heavily rely on agrochemicals. And it is directly linked to the rising levels of obesity throughout the U.S.

In fact, the national problems with obesity can be traced back to the 1970s when the government began subsidizing farmers who grew corn. Cheap industrial corn meant cheaper, high-calorie food. Some of that corn got turned into animal feed, fattening factory farm cattle and lowering the cost of beef, contributing to the excessive amounts we consume today. Unless you buy organic free-range animal products, almost all your eggs, dairy, and meat was raised on subsidized corn. It's also worth noting that corn-fed animal products are much higher in saturated fats—and our increased intake of saturated fats is directly linked to rising obesity.

"Fattening" and "fats" are key words here, as we talk about all this cheap, excess corn. One of the most common corn products is high-fructose corn syrup, which now accounts for 20 percent of the daily calorie intake of many children. Government-subsidized corn syrup is the main reason soft drink and fast food companies make such a huge profit. In the early 1970s the average soda container was a slender eight-ounce bottle; nowadays it's more likely to be a twenty-ounce tub. From bottles of ketchup to children's breakfast cereal, at least one quarter of all processed foods in the grocery stores contain high-fructose corn syrup.

As the government issues stern warnings about rising obesity it continues to support an agricultural policy that makes some of the emptiest and most fattening calories the cheapest and most readily available.

JUNK FOODS, SUGAR, AND VIOLENCE

My little great-nephew Alex (the one who became a vegetarian at four years old) has a horrifying reaction to sugar. Within minutes of consuming even small amounts—in drinks or food—a charming little boy becomes an uncontrollable child, shouting and even hitting people. And it seems that he is one of many, as several of his friends at school are also on sugar-free diets.

My son Grub has a friend who was diagnosed with bipolar personality disorder. As he is a very large man there were times when his behavior was frightening—he actually chased his father around the kitchen table with a knife during one of his manic episodes. It was just after this that Grub read an article about the effect of sugar on the metabolism of some people, and convinced his friend to cut out sugar from his diet. The result was amazing—he became a much calmer person immediately. So I was fascinated to read about a study linking sugar and violence among prisoners.

Stephen J. Schoenthaler, Ph.D., a sociology professor at the Stanislaus campus of California State University, had a hunch there was a link between three alarming statistical curves: the number of incidences of useless violence, the increased consumption of fast foods, and the increased consumption of processed sugars. He convinced a large-scale prison facility in Virginia to help him conduct a study with inmates. Initially the prisoners were fed a typical American diet that included white bread, hamburgers, sausages, fried potatoes, cookies, sweet snacks, and soft drinks. After a few days they were switched to a whole food diet that included plenty of fresh vegetables, fruits, whole meal bread, as well as fish and lean meats.

The results were remarkable. Once they moved on to healthy foods, behavior problems, such as violence and verbal abusiveness, immediately decreased. When they were switched back to the soft drinks and fatty foods, behavior problems returned. The findings created a stir in the national prison network, and Schoenthaler became a much sought after nutritional advisor. He also conducted an interesting study with approximately 8,000 teenagers at nine juvenile correctional facilities. In each site, the standard diet—high in sugar and other refined carbohydrates—was switched to include plenty of fruits, vegetables, and whole grains, as well as vitamins and mineral supplements. During that year, the facilities reported that the incidence of physical violence, verbal abuse, and escape and suicide attempts decreased by almost half.

After twenty years of studying nutrition and behavior at juvenile and adult correctional facilities and in public schools, Schoenthaler is now convinced that the effects of diet are so powerful that everyone should be held responsible for what they eat, just as they are for what they drink when driving. Which means we must do a better and better job of educating people, and ensuring that all these junk foods that we know are causing so many problems are removed from the stores and from our kitchen cupboards.

WASTE NOT, WANT NOT

One of the main threats to our health and the health of our planet is overconsumption. More than a billion of the most poverty-stricken people in the world are suffering and in some cases dying from lack of enough food. Meanwhile, it's estimated that a billion of the wealthiest people in the world are at risk of debilitating diseases and death partly because they are eating too much of the wrong kinds of food.

A person living in an industrialized country consumes, on average, twice as much grain, three times as much meat, nine times as much paper, and eleven times as much gasoline as someone in a developing country, according to the World Resources Institute. With all this consumption comes a huge amount of waste. Timothy W. Jones, an anthropologist at the University of Arizona, spent ten years studying food waste, examining farms and orchards, warehouses and retail outlets, dining rooms and landfills. His research shows that an average family of four currently tosses out $590 per year in meat, fruits, vegetables, and grain products. Nationwide, he says, household food waste alone adds up to $43 billion.

When I first returned from Tanzania, having experienced true poverty firsthand, to the so-called developed world, the thing that utterly shocked me was the waste. The waste of packaging. The throw-away this, that, and the other. And the waste of food. That was the greatest shock of all. The size of the portions of food served, especially in America. The amount left over in restaurants, after school meals, in the homes I visited. And the reason for my distress was not only that I had just spent many years living among people who had almost nothing, but also because I can never forget the lean war years of my childhood. We were taught that waste was one of the greatest sins.

WHAT YOU CAN DO

Tempting and convenient as it can be, we should all avoid fast foods whenever possible and I certainly have given you plenty of incentive for doing so. Each one of us can also make a concerted effort to create as little waste as possible, each day. We can remember to start with smaller piles of food on our plates and ask for reduced portions at social gatherings and restaurants. We can always order more food or go back for seconds. Why not begin small and keep adding if you are still hungry?

When we bring groups of students together for a Roots & Shoots summit, we provide buckets for recycling various waste products, including food. After the first meal we carefully weigh this bucket of food scraps. The kids are horrified when they realize how much they have put on their plates, only to throw it out afterward. The amount of waste is then equated to how long that amount of food would feed a family living in poverty.

If you have a garden, you can make compost from food waste, as we do at my home in Bournemouth. But you can compost even if you do not have a garden. I met a group of youngsters from South Central Los Angeles, in the inner city, who collected food scraps from local residents, and broke it down by using worms. This is known as vermiculture—which can actually be located in the living room for there is no smell! Moreover, it gives the very best quality soil, which the children sold to local parks and greenhouses. They had created a flourishing business!

Chapter 17 The Looming Water Crisis

"We never know the worth of water till the well is dry."

—ENGLISH PROVERB

It may be the worst nightmare of the century, and it is assuming terrifying proportions—as human populations continue to grow we are running out of water, especially clean, safe drinking water. Deforestation around watershed areas is silting rivers with mud washed down from the treeless soil during the rains. And, even as rivers shrink, there are ever increasing demands for the use of stream and river water for irrigation and domestic use. Technology for pumping up water from ever deeper in the earth is decreasing the volume of water in the great underground aquifers. The present heavy use of water for growing corn and soybean mainly to feed cattle—who require additional water themselves—is putting an unsustainable demand on our water resources in many parts of the world. The volume of water in the lower stretches of many rivers is also considerably decreased by the number of dams and reservoirs that have been constructed. And some areas are deprived of water altogether after their former supply has been diverted elsewhere.

I know what it is like to live with severe water shortages. When I first went to Africa, I spent a glorious three months working with Louis and Mary Leakey in Olduvai, that deep gorge that runs across part of the Serengeti plains. The expedition had to run on a shoestring budget. The nearest drinking water was miles away across uninhabited country and in those days there was no road, not even a trail, to Olduvai. The water trailer went to collect water once a week, which meant we were strictly rationed. There was enough to drink—though not to the extent that most Americans consume water today. We had our tea and coffee, and perhaps two or three glasses of water a day, certainly no more. And we were allowed only a quarter of a mug of water per day for washing! Only when we could actually see the car returning with the next week's supply of water could we use what remained from the previous week to have a bath—we had a tiny canvas tub, attached to a folding frame at each corner, into which we each carefully poured our share. Mostly there was not enough for washing our hair—but we found that the Olduvai dust did for us what a dust bath does for certain kinds of birds. And, after all, chimpanzees never wash their hair!

In Dar es Salaam we have had problems with water for many years now. For days on end we had no water in the house because the water pressure was so low it could not fill our tank. We were lucky, for there was a tap outside from which we could collect water in buckets—and we could collect seawater to flush the toilet. Many houses in town did not even have this luxury. This was because of deforestation right along the river that supplies the city, and the increased number of people living in an area that was previously only sparsely inhabited. Now there are more people living there than the existing natural water supply can service. And, as described in the sidebar, the system is old and leaky.

Today we buy water in Dar es Salaam. It comes in a tanker, and pumps water up into the tanks we have constructed at roof level. This water cannot be drunk until it has been filtered and boiled—we are supposed to boil it for at least twenty minutes for safety, but since we have electricity problems as well we seldom cook it that long. In other words, safe water is an expensive and valued commodity.

I constantly think about those people—an estimated 1.2 billion—living with no access to safe drinking water, and a long way from water of any kind. And of the millions who have no access to plentiful firewood with which to boil contaminated water; no wonder their babies get sick. The mothers themselves are so often sick and tired, battered into a sort of lethargic automaton-like condition, and needing all their strength to get through the chores of each day.

WASTING WATER

So I feel sad and angry when I see the terrible waste of water in affluent societies around the world. Next time you go to a restaurant in America, for example, notice how many glasses, partially or even completely filled with undrunk water, are left on the tables when guests leave. Notice how waiters patrol endlessly to refill glasses, even after one has taken but a few sips. In Europe the waiters ask if you want water. They do not refill without asking again. And it is the same throughout the developing world.

When I lecture there is typically a bottle of water at the podium. Usually some is poured into a glass. So I hold it up and ask the audience, "What will happen to the water if I do not drink it?" It will probably be thrown out—without a thought. Even the water remaining in the bottle is likely to be binned since the seal has been broken, although just occasionally people will at least give it to a thirsty plant. I always make a point of holding my bottle up, then taking it with me as I leave the platform.

THE TRUTH ABOUT BOTTLED WATER

Which brings us to a little discussion about bottled water, most of which I learned from a *Seattle Times* article written by a Seattle author, Joshua Ortega (his interesting Web site is www.joshuaortega.com). Most people believe that the best way to be sure that our drinking water is clean and pure is to buy it in a sealed bottle—those that can afford it, that is. But in fact, only a few brands are truly pure. A four-year study by the National Resources Defense Council, released in 1999, found that one fifth of the bottled waters sampled contained known neurotoxins and carcinogens such as styrene, toluene, and xylene. A second study found that, out of 103 brands of bottled water, one third contained traces of arsenic and *E. coli*.

Bottled water, I learned, is one of the developed world's least-regulated industries. Tap water is a public resource, so local authorities must be prepared to provide extensive documentation on its quality to the consumer. But one cannot get access to information concerning the quality of bottled water. It may seem safe, but remember the worldwide recall of Perrier in the early 1990s? This was because it was found to be contaminated with benzene, a poison that has produced cancer in lab animals.

AND ABOUT THOSE PLASTIC BOTTLES . . .

And there is another problem that Ortega points out, which I was certainly not aware of—the production of bottled water is damaging to the environment. The most common plastic used in water bottle manufacturing is PET (polyethylene terephthalate), an environmentally unfriendly substance the production of which produces numerous harmful by-products. Moreover it requires 17.5 kilograms of water to produce only 1 kilogram of PET. In fact, more water is used to make PET bottles than is actually put into them. The Container Recycling Institute reported that of the 14 billion water bottles that were sold in the U.S. in 2002, only 10 percent were recycled—a massive 90 percent ended up in the trash. That's an extra *12.6 billion* plastic bottles for the landfills: bottles that contained water that was no more—and often less—healthy than tap water.

In the U.K. the Food Commission found that some bottled water had traveled more than 10,000 miles. Some supermarket chains such as Waitrose and Fresh & Wild sell water from Fiji, the label of which actually boasts that its source is "separated by over 2,500 kilometers of the open Pacific from the nearest continent." Since we have plenty of our own water in the U.K., transporting water halfway around the world is not only absurd but also a shocking waste of fossil fuel.

WHEN CORPORATIONS OWN THE WATER

A disturbing trend that most people seem unaware of is that multinational corporations are now interested in taking over global water supplies. *Fortune* magazine comments that: "Water is the best investment sector for the century"; the European Bank for Reconstruction and Development comments: "Water is the last infrastructure frontier for private investors"; and in the *Toronto Globe & Mail* was the quote: "Water is fast becoming a globalized corporate industry."

First cattle, hogs, and poultry, then agricultural crops and their seeds—now our water.

Historically local public sector operators, such as city or county utility companies, have been responsible for water supplies. But now the big corporations are becoming increasingly involved. And recent U.S. laws are making it easier for them to do so. Two French multinational corporations, Suez Lyonnaise des Eaux and Veolia Environnement, have swept into America and, along with other multinationals, including General Electric and Bechtel, are taking control of the country's water. When a corporation delivers water, in whatever form, it does so in order to make profits for its shareholders, not to preserve water quality or affordability, and wherever water has been privatized consumers have suffered. When the French privatized their own water services, customer rates went up 150 percent within a few years. Moreover, as mentioned above, the public no longer has the right to facts concerning water quality. A former director of Suez said, "We are here to make money. Sooner or later the company that invests recoups its investment, which means the customer has to pay for it." These are not the people we want to be in control of our water.

In 1998, the privatized water supply of Sydney, Australia (controlled by Suez), was contaminated with *Cryptosporidium* and *Giardia,* yet the public was not informed when the parasites were first discovered. And when the water protection infrastructure in Ontario, Canada, was privatized, the results were disastrous for many communities. In the small Canadian town of Walkerton, for example, seven people died and more than 200 were sickened from drinking *E. coli*–contaminated water in 2000. In Britain, water corporations have a terrible track record: In an eight-year period, from 1989 to 1997, four large corporations, including Wessex (formerly a subsidiary of Enron), were prosecuted 128 times for various infractions.

Many have argued that water is a basic human right and as such it should never be sold on the open market to the highest bidder, and never subject to pricing as decided by the open market. Nevertheless, at least twelve IMF loans that were granted in 2000 were contingent upon water privatization.

About 5 percent of the world's water is already privatized, but water companies are becoming increasingly wary of taking on contracts in developing countries, partly because of political uncertainty, but also because poor countries have learned to negotiate better deals. But as 40 percent of our global population live in countries with water shortages, and as the U.N. estimates that some 2.7 billion people will face water scarcity by 2025, it is clear that new models for conserving and distributing this most precious of all liquids are desperately needed. For, in many countries, both governments and private companies have failed to serve the people, especially the poor.

Anger over Private Water Deals

Dar es Salaam is one of the fastest growing cities in the world, and its privatized water system, installed in the 1950s, has utterly failed to provide for the roughly three million people living there today. There are only some 60,000 people connected to the mains, and it is estimated that two thirds of the available water leaks from broken pipes or is stolen. The poor who have no access to water pay about 12 TZ shillings (about 1.2 U.S. cents) for one liter—twice as much as it costs when delivered to private houses in bulk.

The result, according to Rose Mushi, director of ActionAid Tanzania, is that the poor are financially penalized, they become ill, and they remain poor. She was speaking at a press conference in Dar es Salaam at which the minister of water announced that the government had canceled its contract with City Water (the name given to the British company Biwater, working with German engineers), claiming it had failed to deliver promised new pipelines and improved water quality, and that revenues from water had decreased. The company agreed the project was behind schedule, and that no pipes had been installed, but said quality and quantity had improved, and that the government had given wrong data about water supplies.

Thus ended one of the British government's flagship projects intended to be a model for others in developing countries. In fact, the privatization of water by multinational companies is facing increasing difficulties around the world. It is typically opposed fiercely by unions, but often backed by governments. The role of the IMF and the World Bank in lobbying for contracts for their own companies is widely questioned. According to development groups like ActionAid and the World Development Movement, many countries, like Tanzania, were forced to privatize against their will when the IMF demanded economic reforms or when they applied for loans from the World Bank.

Dave Timms, of the World Development Movement, said at a meeting in London that Tanzania had been forced to privatize its water as a condition of international debt forgiveness. "The IMF forced water privatization on one of the poorest countries in the world in order to benefit Western water companies," he said. And Peter Hardcastle, the policy director of the organization, added, "Donor countries like Britain are piling cash into the private sector. The public sector is not seen as an option."

It was in the 1980s, with early indications of a looming water crisis, that the private sector was seen by the World Bank, IMF, and donor governments such as Britain and France as the only way of raising the money needed to solve it. Then a number of international companies rushed to privatize the water of the poor. They negotiated contracts that gave them monopolies for up to thirty years and guaranteed profits of up to 30 to 40 percent. Some of these companies were accused of paying bribes to governments, and ended up in court. Discontent grew as the water promised to the poor did not materialize, prices increased, jobs were lost, and resentment grew against those private monopolies.

During the past ten years water privatization plans have faced mounting opposition in Trinidad, Argentina, South Africa, and the Philippines. In Ghana, following demonstrations and accusations of high-level corruption, the World Bank eventually withdrew. In Bolivia there have been many protests as a result of privatization. In 2000, in Cochabamba, several people were killed in riots following an increase in water prices and the French company was forced to leave. Early in 2005 another company (a joint venture between the British contractor United Utilities and America's Bechtel) was forced to leave when it tried to increase water prices, claiming the increase was because the Bolivian government had imposed on it a multimillion-pound investment program.

WILL WARS BE FOUGHT OVER WATER?

"Wars of this century will be about water." This was a comment made in 1999 by Ismail Serageldin, former vice president of the World Bank. Most people will agree that the driving economic force behind today's wars is oil. A war over water would be a hundred times worse. Gasoline is a luxury, water is a necessity. If there is one cause in the whole world that crosses all social, national, racial, and economic lines, it's water. And history is rife with conflicts over one party or another's control of a limited resource. This may be one of the most important issues we will face in our lifetime.

WHAT YOU CAN DO

For one thing we can think about water in a different light—as a valuable and increasingly threatened resource. We can stop taking it for granted and stop wasting it. Even little things, like not leaving the tap running while brushing our teeth, will, if enough of us bother, make a big difference. Let us stop throwing it down the drain. Think twice about the amount we use for cleaning and so on. Many toilets use twice the amount of water necessary for a flush; you can get a little gadget to put in the tank that takes care of that problem.

And there is terrible waste of water when it comes to using ice: ice that you use to keep drinks cool, that you pick up at gas stations, that you get free from ice-making machines in hotels, that comes in ice buckets cradling a bottle of champagne or white wine—or even a bottle of water. Ice that you put in a cold box to keep drinks cool on a picnic. What happens to it once it has done its job? Do you even pause to think about the people with no clean water to drink? Remember, too, all the energy that was used to make the ice in the first place.

If you are starting a garden, think about the amount of water available naturally, and take that into consideration when you are deciding what to plant. When you are watering, only do so in the evening, after the heat of the day. Otherwise most of the precious water evaporates.

One of the really important things you can do is to buy a filter for your tap so that you no longer need to buy bottled water—it is only produced and marketed because there is a demand for it. Tap filters are sold at many hardware stores or Internet sites. When local water is controlled by public utilities rather than corporate interests, it is often well regulated and frequently tested for impurities, so it is actually much safer and more reliable than bottled water. Fresh, filtered tap water also tastes just as good, if not better, than bottled water.

If you want to do more than that, then write to your local elected politicians—your senator, congressional representative, or member of Parliament—or better still go and visit them, and tell them that you object to selling your water to corporate interests, especially foreign multinationals. You can also find out about and support public sector programs that are working toward sustainable water solutions. Learn more about the Blue Planet Project; this group is deeply involved in finding ways to solve the world's looming water crisis—perhaps there is a way you can help.

Above all, just think about your water usage, and don't take water for granted.

Gallons of Water Required to Produce About 2 Pounds of Food (Using conventional agriculture methods)

Food	Gallons
Beef	26,400
Chicken	920
Soya beans	530
Rice	505
Maize (corn)	370
Sorghum	290
Alfalfa	235
Wheat	235
Potatoes	130

Source: D. Pimentel et al., Cornell University, 1997, *Bioscience* 42:97–106. *Converted from metric to U.S. customary measure.*

Chapter 18 The Story of Nebraska

"The land is a mother that never dies."
—MAORI WISDOM

For four years in succession I have visited the Platte River in Nebraska to see the awe-inspiring spectacle of the migration of the sandhill cranes, snow geese, and other water fowl. At its peak there may be up to 12 million birds. The cranes arrive during March and stay for several weeks to feed on the corn left over from the harvest in the fields of the Platte Valley. They must get enough body fat for their long migrations north, some going as far as Siberia. For tens of thousands of years this annual migration was supported by the grains provided by the miles of rolling prairies on either side of the Platte, and by the wet meadow complex bordering the river that provided protein in the form of frogs, worms, and so forth. Every night, during their stay, the cranes and geese have roosted along the sandbanks of the beautiful, braided Platte River, which flows above the underlying Ogallala Aquifer—the longest aquifer in the world.

I fell in love with the prairies and the pure magic of the migration, and I have become fascinated by the history of the area, by the tragedy, heroism—and newfound hope—of human interaction with nature in what was once referred to as the Great American Desert. It is a story of hard work, dogged persistence, technological innovation, and heroic effort to produce food from seemingly barren land. And it is also a story of the power of giant corporations pitted against family farms, illustrating what has happened, and is still happening, in so many parts of the world. And I am using this story toward the end of our book because it brings together so many of the themes we have discussed in preceding chapters. And, too, it illustrates the indomitable human spirit and the resilience of nature.

In the thirteenth and fourteenth centuries, before the advent of the white man, the Pawnee tribes cultivated land near the river. They grew ten varieties of corn, seven kinds of pumpkins and squashes, and eight kinds of beans. They did not irrigate and scarcely interfered with the natural environment. And things did not change significantly until the 1860s with the passing of the Homestead Act, which enabled pioneer white families to take many acres of land, free, if they agreed to cultivate it. Despite the challenges that such an undertaking posed, many moved to Nebraska to take advantage of the offer.

This would have a devastating effect on the natural vegetation and wildlife of the prairies; and on the last of the Pawnee Indians, whose numbers had been reduced again and again by diseases brought by the newcomers, and who were being increasingly attacked by the Dakota. By 1875 the last of the Pawnee, resented by the settlers, and with the vast buffalo herds they had depended on for hundreds of years almost eliminated, were forced to move to the "Indian Territory" in Oklahoma.

Meanwhile, the early settlers, despite great hardship, slowly began to tame the prairies. Initially surface water was channeled through ditches in order to irrigate the fields. This was used mostly to water gardens, trees, lawns, and some hay meadows, but also for the growing of grain crops. Then, during the early 1890s, severe drought conditions brought about heavy crop losses—because, of course, when water was not flowing freely in the rivers or streams, farmers could not irrigate when they most needed to. This was when farmers began to search for ways to make use of the fresh clean water of the Ogallala Aquifer, and the first artesian wells were sunk.

It was not until the 1940s and 1950s, with the advent of the turbine engine, that groundwater was truly exploited, for then new and powerful pumps could lift the water from much deeper down. Millions of acres gradually became irrigated as more and more wells were sunk, especially during the drought of 1953–57. (There were fewer than one thousand wells in 1953, 4,500 in 1956. And by 2000 there were 81,112 pumps.)

But the technology that has had the greatest impact on the prairies is the sprawling center pivot, responsible for the huge circular fields so often seen from airplanes, and which can even be seen from space. It has been said that the advent of the center pivot was the most significant mechanical innovation in farming since the replacement of draft animals by tractors. The system consists of a line of sprinklers rotating around a central pivot point, which can creep over hills and through valleys, sweeping over vast patches of land and spreading water where nature is otherwise more conservative. And because the pivots often reach deep down into the earth and have engines that pull the water from the aquifer below, the cumulative effect is draining the water table at alarming and largely unmonitored rates. As more and more center pivots have been installed, it has become possible to cultivate millions of acres of previously unarable land. In 1960, two million acres had been plowed; by 2000 this had increased to 8.4 million.

THEY HAVE THE CORN

Today, the agricultural system in central Nebraska is considered one of the most productive in the world. But it produces mostly the same crop—corn. Nebraska is the third largest producer of corn, after Illinois and Iowa, much of it grown for cattle feed. The next most frequently grown crop is soybeans. The fertile soil of the Platte Valley can also grow other crops, such as sugar beets, dry edible beans, sorghum, wheat, and alfalfa. But because corn is the most responsive to applications of water and fertilizer, some fields have produced only corn for the past thirty years.

There are problems associated with growing this single crop over long periods. For one thing the lack of crop rotation means the potential for loss of soil as a result of wind erosion. For another, it encourages large populations of corn rootworms, corn borers, red spider mites, and others, and these gradually build up resistance to pesticides—requiring ever more chemicals to be used in attempts to eradicate them. These chemicals run off into the surface water, and seep into the groundwater. The aquifer has become increasingly poisoned. Also, with no alternative harvests, farmers suffer terrible losses when the price of corn drops. And this can be manipulated by the big corporations flooding the market, seeking to buy up and control ever more land as small farmers give up the struggle to survive.

THEY HAVE THE FACTORY FARMS

A facility with 20,000 hogs creates as much sewage as 80,000 people—nearly one-third the population of Lincoln. But the distribution of feces produces more than just smell pollution; it also introduces bacteria and toxins into the air and groundwater and—because water sources are invariably linked—also to rivers and streams, where there can be devastating consequences to the local ecology. But all too often, by the time the effects of waste from these facilities are detected, it is too late for local plants, animals, and sometimes humans.

I met a man in Nebraska this spring who had lived and farmed in a small town since 1937. He remembered well the Nebraska hog farms that served an important role in their communities. Legislation for the dumping of hog waste was irrelevant then, because no hog farmer had more hogs than his property could carry. No one would have considered testing the groundwater for contaminants because no one was getting sick—adults weren't developing respiratory problems and babies weren't developing "blue baby syndrome"—and millions of fish weren't turning up dead in their streams.

But things began to change in the 1980s, he said, when big business came onto the scene and the first factory farms, or Confined Animal Feeding Operations, for pigs were established. As these began to raise large numbers of pigs that grew fast, due to hormones and antibiotics administered in the feed, the price per hog quickly plummeted from about $36 to $8. Local pig farmers found it difficult to make ends meet, and many went out of business.

Quite apart from the shocking cruelty of these factory farms, there is the problem of the disposal of pigs' waste. As noted in Chapter 6, this is dumped in open "lagoons." Sometimes it is diluted and sprayed on adjacent (and often genetically modified) crops—fertilizer laced with residual antibiotics and often antibiotic-resistant bacteria. The stench of this waste is revolting. My friend Tom Mangelsen has his permanent residence in Jackson Hole, Wyoming, but he still maintains a cabin in Nebraska, the state in which he was raised. On one of my visits to Tom's cabin the nearby field had just been treated with the diluted pig waste, and the smell made us both feel quite sick. No wonder the value of properties drops when a hog operation is built within smelling distance. The dung of the pigs on the farms of my childhood did not smell like that, not in the least. No wonder the American pork industry targets poor rural towns where it is cheapest to win political favor and easiest to silence critics.

"You could never dump this stuff in New York City," said Laura Krebsbach, one of the nation's experts on state and federal regulations pertaining to Confined Animal Feeding Operations, "but it's hard to get a big outcry in Cherry County, Nebraska, where the population is one person per square mile."

THEY HAVE THE GLOBAL SUPERMARKET

Farmers here face many other problems. In his insightful book *Eat Here*, Brian Halweil explains how, as recently as the 1950s, virtually all of the fruits and vegetables consumed in Nebraska cities were grown locally. Then long-distance transport of food became possible with refrigerated long-haul trucks, cheap gasoline, advances in food processing that made long-term storage possible, and a federally subsidized interstate highway system. This heralded the arrival of huge supermarkets appearing in more and more cities and towns selling food from all over the country and from overseas. And so, as local farmland declined in profitability, thousands of family farms in Nebraska (and the surrounding states) went under and farming communities were increasingly replaced by subdivisions and concrete. And the remaining farmers mostly depend on good harvests of corn and soybeans for their survival.

The Visionary Nebraskan

George W. Norris (1861–1944) started life farming in Ohio, on his family's farm, and was always a friend to the farmer and the common man. He moved to Nebraska at twenty-four to study law, fell in love with the land, and was eventually elected to Congress where he became an extremely effective senator, and the most prominent political figure in Nebraska's history.

Norris's visionary concept was to combine flood control, irrigation, groundwater recharge, in-stream flow maintenance, land reclamation, reforestation, and electric power generation into a single giant program. His efforts to make this project a reality were frustrated for twelve years by a major controversy—namely, whether it should be developed by private companies or by the federal government. Congress at first seemed to assume that private development was the only option, but Norris campaigned tirelessly for government management of water resources. He believed water to be a God-given natural resource intended for use by the people, not a commodity to be exploited for profit. In his autobiography he wrote, "From the beginning to the end, there was that irreconcilable conflict between those who believed the natural wealth of the United States can best be developed by private capital and enterprise, and those who believe that in certain activities related to the natural resources only the great strength of the federal government itself can perform this most necessary task in the spirit of unselfishness for the greatest good to the greatest number." How shocked he would be to know that the "great strength of the federal government" is now being used to strengthen the power of private capital and enterprise to the detriment of the small family farmers for whom he worked so tirelessly. And he would be even more shocked by the power of today's multinational corporations and their bid to take over global water supplies.

AND THEY HAVE THE WATER CRISIS

In the previous chapter we discussed the looming global water crisis—nowhere is this illustrated better than in Nebraska and the surrounding states. At the present time (spring 2005), Colorado, Nebraska, and Wyoming are in the fifth year of a drought cycle and there is growing concern about sinking levels in both surface water and the underlying aquifer. Water is used not only for agriculture, but increasingly for domestic use in fast growing towns. Already, in some parts of Nebraska, moratoriums on drilling new wells have been introduced, and plans to impose similar regulations in other parts have led to a frenzy of well drilling—indeed the pace of drilling has been more than a match for drilling restrictions—the number of new wells in the last three years accounts for more than 44 percent of the overall total for the last ten years. In 2004, for the third year in succession, the statewide total of new wells topped 1,000. So far, there are no restrictions on the amount that can be pumped, but this is likely to change, as it has in other states. The major problem is that both corn and soybeans require a great deal of water—in 2003 alone farmers flooded 8.5 million acres to a depth of one foot. Some of this was drawn from surface water, but most was pumped from underground sources.

I first learned that the Platte River was endangered in 2003. This was such a horrifying thought that I asked Tom if he could arrange for me to meet with some of the local farmers and conservation groups. The following year, thanks to the efforts of Tom's friend Paul Johnsguard, we attended such a meeting and heard about the problems that faced both groups. At first the farmers were reluctant to talk, but once they had relaxed it became clear that they were facing tough times. They felt threatened by the big corporations that could control corn prices. Each farmer, to make ends meet, was working an area that used to be divided into four or five farms. They were plowing their land right up to the road, destroying the last bits of natural habitat for wildlife, because every extra foot counted. And they resented the visitors who came once a year to watch the crane migration and just walked over their land without a by-your-leave or thank-you. They also resented the conservation groups that bought up land, because they paid inflated prices and this drove up the value of all land in the area so that the farmers then had to pay increased taxes.

I told them about the problems faced by the farmers around Gombe, and some of the ways in which we tried to help them so that they, in turn, would help us care for the chimps. How could we help these farmers in the Platte Valley, I asked. We discussed the various solutions that have been proposed.

First, there are conservation easements. If a farmer signs on to this program he agrees not to develop certain parts of his land, and not to allow developments on it, in perpetuity. This entitles him to a variety of tax breaks. Another scheme pays farmers along the Republican and Platte Rivers, all the way from the Wyoming border, not to irrigate. This is designed to preserve water and return some 100,000 acres of cropland to grass. The program began in 2002 and so far 19,818 acres have been protected. All the farmers said they would certainly take advantage of this offer, but were skeptical that the money would be forthcoming. And there were a few small schemes where conservation groups paid farmers not to harvest their crops, thus leaving more grain for the birds.

One old farmer began to talk about the old times. He remembered waking in the night as a child to the roar and thunder of the Platte River in spate. The spring melt of snow in Colorado sent the water racing down from the mountains, clearing out the sand and silt and creating new channels. "Too many dams and reservoirs now," he supposed. And he remembered drinking water from the family well when he was a child. Cups full of clean, cold water. "I won't let my grandchildren drink even one teaspoonful of that water now," he said. "It's all those chemicals we've put on the fields for the corn that's done it."

He lapsed into silence. Thinking, I suppose, of a long-gone world. A world that was clean. A world that was hard for a farmer, but one in which he worked in the expectation that his children would inherit his land, and his love of the land. Now, the farmers told us, their children seldom stayed. They went to seek their fortune in the towns, away from the backbreaking toil, the struggle to make ends meet, the shrinking river that flows over the sinking, poisoned aquifer.

And we talked about the cranes and the other birds. But for the farmers the cranes would be in a sorry plight. The cranes depend on the grain left on the ground after the harvest to give them sufficient body fat to migrate onward to Alaska and Siberia. This dependency has been caused by the destruction of the prairie and the animal protein of the wet meadows. Perhaps there is some way that the cranes, and the visitors they attract, can help the farmers as they struggle to survive. Some of the farmers have built little viewing blinds when their land is adjacent to the river, places where the increasing number of tourists can view the cranes arriving in the evenings.

YET HOPE GROWS IN NEBRASKA

What we did not discuss, when I met with the farmers, was the glimmer of hope offered by the Slow Food movement, known by some as the food revolution. Before I began work on this book, I had never really thought about it. But the Slow Food movement has taken hold in Nebraska.

When I visited Lincoln in 2004 to give a talk at the university there, I had not heard of John Ellis and his Centerville Farmers Market—a store that he opened with other local farmers. It was stocked almost entirely with food that represented fifty farmers from within fifty miles of Lincoln. It was a brand-new, revolutionary idea when he first sold his farm and equipment to start it up. I read about John Ellis in Brian Halweil's *Eat Here*. The venture epitomizes the resilience of the human spirit—in this case linked to the resilience of nature.

John started his store even though he was well aware that there was a Wal-Mart Supercenter on the other side of town, with twenty-eight aisles of food. He believed that customers would buy from the Centerville Farmers Market because they wanted locally grown food. And in the Wal-Mart he knew that almost all the food had traveled thousands of miles to get there. Even a lettuce farmer just outside Lincoln must, if he wants to sell to Wal-Mart, send his produce 225 miles to North Platte to be inspected (each lettuce must comply with strict regulations concerning quality control and appearance). After which it is sent 225 miles back to be sold in the store in Lincoln.

Returning Bison to the Prairies

Even as agribusiness corporations are acquiring more and more land from which more and more profits can be squeezed in the short term, there are some who are acquiring land with more sustainable business models in mind.

One such pioneer is famed media mogul and naturalist Ted Turner. When I met him in New York recently, we discussed his vision. In Nebraska alone, Ted owns five ranches that cover nearly 388,000 acres (about 600 square miles). Ted aims to "manage his lands in an economically sustainable and ecologically sensitive manner while promoting the conservation of native species." The implementation of these objectives is proving, on all his ranches, that ranching along with limited, sustainable timber harvesting and recreational opportunities can be economically viable.

Ted Turner has also dedicated significant resources to rebuilding the natural ecology of his properties. The prairie has been revitalized and wetlands, once drained for hay production, have been restored. The bison graze on natural prairie grasses and other native wildlife is protected. In all, Ted Turner has acquired a total of two million acres of land in the U.S. and has focused his conservation efforts on nearly two dozen wild animal species—many of which are listed as threatened, endangered, or on the brink of extinction. He has thus demonstrated to the farming community that it is possible to make a profit from ranching while, at the same time, restoring and preserving the beauty of the prairies. What a great example of a highly successful businessman who is using his wealth to make this a better world.

In North Platte, the regional distribution center for Wal-Mart comprises a series of sprawling, hangar-sized warehouses containing the giant freezers, ripening rooms, and packing stations for all the "fresh" food—vegetables, fruit, meat, milk, and so on—that will be sold in the giant supermarkets throughout the whole of the American Great Plains region. How could John Ellis hope to compete? Ultimately, he couldn't. The Centerville Market closed in the fall of 2004. Nonetheless, Ellis' passion and efforts raised consumer awareness. Since then, a Slow Food Nebraska chapter was launched in the winter of 2005. In May 2006 a statewide "Buy Fresh, Buy Local" resource guide began listing the many restaurants, institutions, and groceries that use and/or stock farm goods raised within the region.

Meanwhile, Open Harvest, the food cooperative in Lincoln, is supplying the growing demand for local foods. Many other such stores are springing up. And more and more farmers markets are appearing in Lincoln and surrounding areas. Farmers are banding together, so that more people can be supplied with a greater variety of produce. There is growing support among politicians and voters for measures such as conservation easements and tax credits that will protect the farmers and their land—measures that also help to preserve the beauty and biodiversity of the countryside. Many farmers markets have become venues for social gatherings, music, and exchange of news, adding to their popularity. People get to know the farmers who grow their food, and sense that these farmers will be more responsible in their production methods.

Within the city limits of Lincoln is Shadowbrook, a thirty-six-hectare certified organic farm that serves seventy families. Community gardens are springing up here, as in so many places around the world, where city people can find the inner peace that comes from working with the soil, marveling at the growth of life that they have planted, hearing the song of birds and the humming of the honeybees. The harvests from these community gardens, and those of the farmers growing for their local communities, are truly harvests for hope. And the more people who support these efforts, who use their hard-earned dollars to buy produce from farmers markets, and their political power to vote for measures to protect the source of this produce, the more crop diversity will return to the despoiled land. Nebraska is even growing grapes and making its own wine today.

On the last evening of my stay there was one of the most amazing sunsets I have ever seen. Palest shell pink turning an unreal crimson, then purple, then fiery red. Tom, my sister, and I sat on a rise overlooking the river and watched and listened as the cranes came in, skein after skein. Their wild voices filled the air, a great volume of sound drowning out the trucks on the interstate. There was still enough food, still enough water, to sustain this ancient migration. Somehow, together, we must ensure that our grandchildren are not denied the opportunity to marvel, in their turn, at one of nature's most spectacular scenes. The last glow of the sunset was still smoldering against the almost dark sky as we walked back to the cabin, silent with the magic of the evening. The pale stubble of the corn harvest stretched away from the river into the distance; the dark shape of a center pivot loomed against the sky; a truck thundered along the highway. And the cranes were still calling.

One Man's Memories

By photographer Tom Mangelsen, who grew up on the prairies of Nebraska.

"From a twelve-year-old's perspective, Ned Martin was a giant of a man, over six feet tall with bright blue eyes; always in overalls and a sweat-stained and wrinkled straw cowboy hat covering his balding head. The hat seemed almost as permanent as his smile.

"In the corner of his mouth there was always a toothpick, ready for the next steak I suppose. Nearly everything we ate was raised on Ned's ranch, mostly hogs, chickens, and cattle. Ned was a proud and gentle soul who cared about his livestock. He often named his favorite animals. They were more to him than just another dollar or another pork chop. The chickens ran free. The henhouse had kerosene lamps that provided a measure of warmth on colder nights. The barn was full of horses and milk cows. Mostly, Ned's animals had space; they had a life beyond being fattened for the market. Ned also loved having wildlife around his place. He left brushy hedgerows and woodlots for pheasants, rabbits, and deer. Red-tailed hawks sat in the cottonwoods. He saw no need to have the tidiest ranch in the neighborhood and most of Ned's neighbors felt the same way. There was plenty of room for both wildlife and livestock.

"The scene in front of me now is a far cry from what I remember. Wading knee-deep in a quagmire of olive brown excrement and mud, the herd inches forward. They stretch their necks toward the prone stranger's camera lens, nostrils flaring, with curious black liquid eyes nearly the size of baseballs. For the first time in my life I have come face-to-face with all those hamburgers and T-bones I have eaten and suddenly feel ill. The feedlot is huge, a mile square. The closest cattle are Black Angus, but beyond them are several thousand white-faced Herefords, some standing on the drier mounds that have been pushed up into small hills by bulldozers, the majority wading in the muck. Earlier in March it had rained and snowed for days and I imagine the scene—far more miserable. Fortunately it has been warm and windy the past week. Even with the wind at my back, the acrid smell is barely tolerable. Wet and mildewed, the grain next to the feedlot had been ground up with cornstalks into a silage mash smelling of molasses. Not a woodlot or brushy hedgerow is in sight, and only an occasional cottonwood. This feedlot is one of the largest of a dozen or so in the area, a few miles west of Grand Island, Nebraska, bordering the old ordinance plant where bombs and munitions were stored during World War II. This location was chosen because it is in the center of America, far from any shore.

"So how did it happen, this scene before me of cattle knee-deep in their own waste? Grain storage elevators overflowing and mountains of government-subsidized surplus corn piled on the ground?

"Fifty years ago much of the rolling prairie grasslands remained intact, impossible to irrigate because of the topography. Today only the steepest of hills and gullies remain untouched, thanks to the advent of the pivot irrigation sprinklers crawling over the mostly barren landscape where woodlands and native prairie used to be. It is true a few wildlife species have achieved some benefit from modern farming technology and the abundance of grain. Migrating waterfowl and cranes feed on the grain left from the harvest that fuels their migration to the far north. And with most predators, like coyotes and cougars, having been killed off, wild turkeys and white-tailed deer have become more prevalent. However, many species have been devastated. Formerly abundant, cottontails, prairie chickens, and sharp-tailed grouse have nearly disappeared. In the alfalfa field where I counted 123 jackrabbits as a child one summer night, I haven't seen a jackrabbit in more than thirty years.

"My fondest childhood memories were those summers spent on Ned Martin's place east of Maxwell, Nebraska, and on the Schneider farm near the Platte River where we still have our cabin. The Martins, although they grew some crops, liked the idea of being more rancher than farmer. They ran cattle in the short native prairie grasses of the Nebraska Sandhills. Back then the morning sounds of early spring were enriched by the crisp, sharp calls of meadowlarks and song sparrows. The prairie-scented air, which was always cool, made one breathe deeply. Leaving their leks, or dancing grounds, flocks of low-flying prairie chickens and sharp-tailed grouse flew over the valleys in search of morning feed.

"The sight of cattle in those days, spread out over the vast grassland, seemed healthy and good. From time to time cowboys herded the cattle to greener pastures and in winter moved them to areas where they would be fed from haystacks of prairie grass or alfalfa. Their meat was lean and flavorful. Ned Martin showed me the right way to cook a steak, from beef he had raised. Under a large cast-iron frying pan Ned turned the propane burner on high, took a twenty-ounce hand-cut sirloin from the refrigerator, heavily salted it on both sides, flopped it into the pan, counted to five, flipped it over and after another five seconds it was ready to eat. I was impressed.

"It's been nearly three weeks since I was on my stomach eye-to-eye with the herd of corn-fed Nebraska prime. The sign read 'No Trespassing—Bio Secure Area.' I didn't know if the sign was for my protection or the cattle's. I was careful not to trespass an inch past the fence. Here in the Heartland it is not just loss of habitat that is responsible for species decline but the amount of chemical fertilizer and herbicide and pesticide poisons put on the fields to create bumper yields of corn. The same corn that feeds the cattle in the feedlots. Industrialized farming has forced many family farms to go the way of the jackrabbits. Ned, too, is gone, but I shall never forget his respect for animals and the land on which he raised them."

Chapter 19 | Harvest for Hope

"If you lose hope, somehow you lose the vitality that keeps life moving, you lose that courage to be, that quality that helps you go on in spite of it all. And so today I still have a dream."

—Martin Luther King, Jr.

We live in troubling times. The giant corporations control much of the world's food, as well as the patents on our seeds. Billions of farm animals live in conditions of utmost deprivation and misery. Humans and animals are increasingly becoming poisoned from the chemicals that have been lavishly sprinkled over fields, crops, and food produce and that have contaminated the earth's water, soil, and air. Disease-causing bacteria are building up resistance to the antibiotics that are routinely administered to livestock in factory farms. Genetically modified organisms, GMOs, have escaped into the environment and who knows what that will mean? Billions of tons of fossil fuel are used to transport our food from one end of the planet to the other—and often back again—contributing significantly to the changes in global climate. And the soil of our planet is being not only poisoned but swept away by the wind from areas cleared for agriculture. Monoculture crops subsidized by governments provide fuel for the manufacture of hamburgers and T-bone steaks. Thousands of children die of obesity and its attendant ills in the West, while millions more die of starvation in the developing word. Family farms are going out of business and asphalt and concrete is spread over more and more good arable land. Water is becoming terrifyingly scarce as well as polluted.

All this and more makes grim reading, and while I was working on this book I had nightmares as I learned more and more about the unethical conduct of some of the largest multinational corporations. They are so powerful, and they can break down those who oppose their will with lawsuits that only they can afford. Many corporations contribute large amounts to the campaigns of politicians: They are repaid by support for their programs. Money and power are falling into the hands of fewer and fewer people on the global stage.

In 2005 the United Nations issued a rather daunting "Millennium Report." After a five-year study, a team of international scientists came to a sobering agreement: Unless we stop the pollution and degradation caused by industrial farming and seriously address overfishing and global warming, we will literally run out of enough resources to feed everyone by the year 2050. The scientists used the analogy of people overspending their bank accounts. Meaning, to put it bluntly, that if governments and industries continue to allow and actually subsidize farming methods that destroy our planet's resources, for the sake of immediate profit, we shall eat everything edible to the point of human population collapse—and we shall take many other species with us.

Fortunately, the report indicated that the situation is not utterly hopeless. Not, that is, if we take steps immediately to reduce fossil fuel emissions; bring to an end government and consumer support for industrial agriculture, including animal factory farms and fisheries that harm the planet; and start subsidizing and supporting more sensible and sustainable ways to feed human beings. "Take immediate steps"—that is addressed to each one of us. That is why I have spent countless hours I did not have working on this book.

There has never been a time when it is more crucial for us to carefully consider where our food is coming from and how it was grown, raised, and harvested—so that we can make informed efforts to purchase the right things. For our choices will affect not only our own health but also the environment and animal welfare. And, too, our choices will affect small family farms. I have told the stories of several farmers who have returned to more traditional farming methods, working—usually very hard indeed—to have their produce branded organic and to become once more wise stewards of the land. It is desperately important that we support them by buying their produce whenever possible. And persuading our friends to do the same.

It is important that we talk, often and enthusiastically, about the *positive* developments that are going on around the world. For one thing, it does seem that more and more people have woken up to what has happened to our food, have begun to comprehend the almost unbelievable mess we have made. And so people are starting to protest. To protest against the insults being perpetrated, in the name of progress, against people, animals, and the environment, and the unsustainable demands that are being made on our children's future, the future of our planet. Some of these protesters took on awesome tasks: like Percy Schmeiser, who took on Monsanto; and the two New York teenagers Jazlyn Bradley and Ashley Pelman who stood up to McDonald's. Of course, there are countless cases when people have fought against injustice and lost. Robert Kennedy won his case against the pork barons, only to have the rug snatched from under his feet when the legislation was subsequently changed to allow them to continue polluting. But every time a stand is made there are more people who hear about what is going on. More people who can work in their own way to make a difference.

I find hope in the people who join forces to buy farmland to save it from development and those who pull genetically modified crops out of the ground. Others organize and operate the farmers markets and cooperatives. Some are involved in the Slow Food movement, and many restaurateurs, such as Alice Waters of Chez Panisse and Tod Murphy of the Farmers Diner, are changing the world one plate at a time. It seldom makes front-page headlines when a company opens an organic food line, when a restaurant chain commissions its produce from a local organic farmer, or a household decides to join a Community Sponsored Agriculture group, a CSA, yet these are the actions that bring me hope, for they are already making change in the world.

There is hope, also, in the resilience of nature, her ability to repair the wounds we have inflicted. TACARE, the Jane Goodall Institute's program to improve the lives of the villagers living around the Gombe National Park, is a perfect example, for innovative techniques have restored life and productivity to land that had been overfarmed and abandoned, most of the topsoil washed away. It is harder, and takes a very long time and very hard work, to rescue land that has been polluted by years of chemical poisons, but, as I have shown in this book, it can be done.

And there is hope in the growing number of people who care, and do something about it. The wonderful mother-daughter team, Frances Moore Lappé and Anna Lappé, in their book, *Hope's Edge,* describe "a new social mentality" that has been developed in Belo Horizonte, the fourth largest city in Brazil. There was a time when one fifth of the children there were malnourished and poverty was rampant. Then, in 1993, Belo became "the only city in the capitalist world that has decided to make *food security a right of citizenship*" (my italics).

The city improved the way the food market worked. It provided all the students in the city's schools *four* nutritious meals per day, with ingredients mostly from local farmers. The city set up produce stands for some forty local farmers. It owns and operates the Restaurante Popular, which serves 6,000 meals a day at less than half the market price. This is all made possible by the twenty-six warehouse-sized stores that sell local produce at fixed prices—often half the price of nearby grocers. These stores are on government-owned prime urban real estate rented out to entrepreneurs at rock-bottom prices—the government reserves the right to set the price, and the vendors must make weekend deliveries to the poor.

There is a Green Basket program that links hospitals, restaurants, and big food buyers to local, organic growers. There is a local food council that helps to form partnerships with church and labor groups, and advises the government on ways of improving the food system. The entire program consumes only 1 percent of the city's budget and it is reckoned to be extremely cost-effective—the children do better in school with the opportunity of being productive citizens when they leave, and the entire population of the city is far healthier.

Indeed, the human brain, this organ—the spongy mass of gooey cells we all house in our skulls—is capable of the most wondrous technologies. Unfortunately, when there is a disconnect between mind and heart, technology can be—and has been and is being—used for evil purposes. Unless our intellect is bonded closely with our feelings of love and compassion, although we may still be very clever, we shall not be wise.

Fortunately, my constant traveling gives me many opportunities to meet many of the wise, such as Dr. Hugo Hubacec and Dr. Peter Kromer, who came to see me in Vienna in the spring of 2005 to explain their remarkable technology. "SIPIN Technology" can revolutionize farming in arid conditions and greatly reduce hunger. Because no sophisticated machinery is involved it can be used easily by the local villagers.

SIPIN is a water-absorbent, natural silicate powder that is adjusted to the local conditions and used in combination with local soils. It is applied in the area around the plant's roots, and covered with local earth or sand. Over time SIPIN turns into amorphous clay minerals and stable, natural clay-soil compounds. These show high water absorption. After treating the roots of one plant, others can be planted in the same place without further applications for at least three years. It is possible to save up to 75 percent of the water otherwise used to sustain the crops, which, in a dry area, can save lives. By using SIPIN, the available water resources can nourish three to four times more people. "We do not want SIPIN to be marketed by one of the big corporations," Dr. Hubacec told me. "This is just to help people." It was an exciting meeting and we plan to work together to take this remarkable technology to places where it is so desperately needed.

CHANGING THE WORLD: ONE PURCHASE, ONE MEAL, ONE BITE AT A TIME

As marketing people tend to do, they have identified a strong force on the consumer landscape—people who value a lifestyle of health and sustainability and are willing to pay for products that support their beliefs. This group, they say, has adopted a Lifestyles of Health and Sustainability—LOHAS. It's estimated that 68 million Americans, about a third of the adult population, qualify as LOHAS. Without realizing it, these people have become the most influential force in the recent food revolution. And they have numerous allies, including farmers who want to create a more healthy and sustainable land, public health officials concerned about toxins and antibiotics in food products, environmentalists who are concerned about pollution from factory farming, consumer rights groups wanting more accurate labeling of food sources on food packaging, and union activists who want safe working conditions for their members who run the risk of massive exposure to toxic pesticides and chemical fertilizers.

Yes, collectively we, the people, are the force that can lead to change. Every time we go shopping for food, every time we chose a meal in a restaurant, our choices—what we buy—will make a difference—not only for our own health and our own peace of mind, but also for the future of the planet. Fortunately more and more people are beginning to realize this. Every time an individual makes such a change in his or her lifestyle the number of people eating ethically and healthily increases—by one.

This philosophy—that every individual matters and every individual makes a difference every day—is at the heart of the Jane Goodall Institute's Roots & Shoots global program for youth. The name is symbolic: Roots make a firm foundation and Shoots seem tiny but to reach the sun they can break through a brick wall. Imagine the brick wall as all the problems we humans have inflicted on the planet. The message is one of hope: Thousands of young people—and there are some 7,500 groups established in more than ninety countries—are breaking through all manner of brick walls to make the world a better place.

In several African countries there are Roots & Shoots groups maintaining tree nurseries and distributing seedlings to schools that are otherwise surrounded by compacted, sunbaked earth. As the trees grow, cared for by the students, grasses can survive in the shade below them. And, encouraged by the greening of the schoolyards, Roots & Shoots groups are growing fruit trees and vegetables to improve their diet. Refugee groups are also growing vegetables, and some keep chickens for their eggs.

There are so many projects and many of them concern food and farming. Students are composting, making organic vegetable plots. Two groups (one in the U.K. and one in Belgium) rescued hens from battery farms, and studied how they grew back their feathers and adjusted to freedom. They are campaigning (always without violence) against the use of synthetic chemicals in food; the feeding of hormones and antibiotics, as prophylaxis, to livestock; the use of pesticides, herbicides, and chemical fertilizers on farmland; the use of nonbiodegradable containers in school lunch boxes. They are writing letters to legislators, raising money for all manner of causes. And they are influencing their parents.

Roots & Shoots groups are thus joined with all of us who are trying to change the world. In fact, many predict that in the years to come, activists will have a greater impact through consumer choices and the culture of economics than through legislative lobbying or lawsuits.

EATING TO VOTE

Remember, every food purchase is a vote. We might be tempted, as individuals, to think that our small actions don't really matter, that one meal can't make a difference. But each meal, each bite of food, has a rich history as to how and where it grew or was raised, how it was harvested. Our purchases, our votes, will determine the way ahead. And thousands upon thousands of votes are needed in favor of the kind of farming practices that will restore health to our planet.

Our world can no longer afford the heedless consumption of the Western world that is now spreading its greedy tentacles around the globe. The price, most of which must be paid by our children, is too great. Only by acting together, by refusing to buy food that has been secretly laced with poisons and pain, can we make a stand against the corporate power that is circling our planet. So let us join hands. Let us speak out for the voiceless and the poor. Let us assert our right, as citizens of free democracies, to take back into our hands the production of our food. Let us, together, sow seeds for a better harvest—a harvest for hope.

Resources

These are just a few of the many resources available for eating mindfully. Visit our Web site at www.harvestforhope.com for a greatly expanded list of resources. The Web site also offers personal messages from Dr. Goodall, as well as news updates and a community forum for sharing information on sustainable foods.

HARVEST FOR HOPE

There are so many resources available for eating mindfully today that this list can by no means be comprehensive, but we hope it serves as another step in your quest to educate yourself and others on responsible nourishment and sustainability.

Take Action with These Organizations

Blue Planet Project—www.blueplanetproject.net
The Blue Planet Project is an international effort begun by the Council of Canadians to protect the world's fresh water from the growing threats of trade and privatization.

Chefs Collaborative—www.chefscollaborative.org
Chefs Collaborative is a national organization that promotes sustainably raised, locally grown, artisan food. Membership is open to anyone interested in the interconnectedness of the environment and food choices.

***Earth Day Network's Footprint Quiz*—**
www.earthday.net/footprint
Earth Day computes sustainability in specific and understandable terms by using the best available scientific data. They allow individuals, policy analysts, and governments to measure and communicate the economic, environmental, distributional, and security impacts of natural resource use. Take the Footprint Quiz to find out how big a footprint you make upon the earth—it's eye-opening.

EarthSave—www.earthsave.org
EarthSave International was founded by celebrated author John Robbins, as the direct result of the overwhelming reader response to his book *Diet for a New America: How Your Food Choices Affect Your Health, Happiness and the Future of Life on Earth.*

True Food Network—http://www.truefoodnow.org
Center for Food Safety, 660 Pennsylvania Ave. SE, Suite 302, Washington, D.C. 20003
Phone: 202-547-9359
The True Food Network offers a valuable resource called the True Food Shopping List. While not a complete manual, it's a starting point for consumers who want to shop smarter and safer, listing many popular brands of food and whether or not they contain GMOs.

The Jane Goodall Institute—www.janegoodall.org
4245 North Fairfax Dr., Suite 600, Arlington, VA 22203
Phone: 703-682-9220
The Jane Goodall Institute is a global nonprofit that empowers people to make a difference for all living things. We are creating healthy ecosystems, promoting sustainable livelihoods, and nurturing new generations of committed, active citizens around the world.

The Monterey Bay Aquarium—www.mbayaq.org
The aquarium's mission is to inspire conservation of the oceans, but this resource-rich Web site has a wealth of facts and useful tools to achieve that goal. A must-have is their free Seafood Watch wallet card (which can be downloaded off their Web site), recommending which seafood to buy or avoid, thus helping consumers to become advocates for environmentally friendly seafood.

New England Heritage Breeds Conservancy—www.nehbc.org
This organization works to conserve historic and endangered breeds of livestock and poultry and encourage production of these breeds to advance farmland.

Organic Consumers Association—www.organicconsumers.org
6771 South Silver Hill Dr., Finland, MN 55603
Phone: 218-226-4164
Representing over 600,000 members, the OCA is a grassroots nonprofit public interest organization that deals with crucial issues of food safety, industrial agriculture, genetic engineering, corporate accountability, and environmental sustainability. It is focused exclusively on representing the views and interests of the estimated ten million organic consumers in the U.S.

Slow Food—www.slowfood.com
Slow Food USA
20 Jay Street, Suite 313, Brooklyn, NY 11201
Phone: 718-260-8000
Founded by Carlo Petrini in Italy in 1986, Slow Food is an open-membership international association that promotes food and wine culture, but it also defends food and agricultural biodiversity worldwide. It includes over 83,000 members worldwide, with offices in Italy, Germany, Switzerland, the U.S., France, Japan, and Great Britain.

Soil Association—www.soilassociation.org
The Soil Association is the U.K.'s leading campaigning and certification organization for organic food and farming. It was founded in 1946 by a group of farmers, scientists, and nutritionists who observed a direct connection between farming practice and plant, animal, human, and environmental health.

Sustainable Table—www.sustainabletable.org
Sustainable Table is a consumer campaign developed by the Global Resource Action Center for the Environment (GRACE). It was launched to help fill in the gaps in the sustainable food movement, and to help direct consumers to the leading organizations working on the issue.

USDA Agricultural Marketing Service—
www.ams.usda.gov/farmersmarkets
The USDA Agricultural Marketing Service offers a state-by-state listing of local farmers markets.

Worldwatch Institute—www.worldwatch.org
Founded in 1974, the Worldwatch Institute offers a unique blend of interdisciplinary research, global focus, and accessible writing that has made it a leading source of information on the interactions among key environmental, social, and economic trends.

COMMUNITY SPONSORED AGRICULTURE (CSAs) AND LOCAL COMMUNITY GARDENS

American Community Gardening Association—
www.communitygarden.org
This is a binational nonprofit membership organization of professionals, volunteers, and supporters working toward community greening in urban and rural communities.

Compost Guide—www.compostguide.com
This clear and comprehensive guide to Everything Compost, including vermicomposting (worm composting), has an extensive list of resources for starting a garden.

Food Routes—www.foodroutes.org
This national nonprofit is dedicated to reintroducing Americans to their food, the seeds it grows from, the farmers who produce it, and the routes that carry it from the fields to our tables. Among their Tools for Advocates is the publication *Where Does Your Food Come From?*, which discusses how to effectively develop a local foods campaign and what kinds of messages resonate with public audiences.

Local Harvest—www.localharvest.com
220 21st Ave., Santa Cruz, CA 95062
Phone: 831-475-8150
Local Harvest maintains a definitive and reliable "living" public nationwide directory of CSAs, farmers markets, small farms, and other local food sources.

Healthy Schools

The Community Food Security Coalition (CFSC)—
www.foodsecurity.org
This excellent Web site provides a Farm to School Program guide, offering tips, tools, technical assistance, and funding opportunities, in addition to another useful publication, *Healthy Farms, Healthy Kids*, among others.

The Edible Schoolyard—www.edibleschoolyard.org
The Edible Schoolyard offers its program as a model for other organizations engaged in creating organic gardening and cooking projects for children.

National Farm to School Program Web site—
www.farmtoschool.org
These programs connect schools with local farms with the objectives of serving healthy meals in school cafeterias, improving student nutrition, providing health and nutrition education opportunities that will last a lifetime, and supporting local small farmers.

USDA Report—www.ams.usda.gov/tmd/mta/publications.htm
Titled "*How Local Farmers and School Food Service Buyers Are Building Alliances*," this report summarizes the educational highlights of the workshop in an effort to help small farmers and school food service buyers throughout the country explore how they might be able to establish similar business relationships in their own communities.

ANIMAL RIGHTS ORGANIZATIONS AND SANCTUARIES

Compassion in World Farming Trusts—www.ciwf.org
CIWF's mission is to work toward ending factory farming systems and all other practices, technologies, and trades that impose suffering on farmed animals, by such means as hard-hitting campaigning, public education, and vigorous political lobbying.

Farm Sanctuary—www.farmsanctuary.org
P.O. Box 150, Watkins Glen, NY 14891
Phone: 607-583-2255
Farm Sanctuary, a national rescue, shelter, and farm animal adoption network, was started in 1986, when two caring activists rescued Hilda, a sheep who had been abandoned and left for dead on a stockyard "deadpile." Today, Farm Sanctuary is the largest farm animal rescue and protection organization in the U.S.

People for the Ethical Treatment of Animals—www.peta.org
Founded in 1980, PETA, with more than 850,000 members, is the largest animal rights organization in the world. PETA is dedicated to establishing and protecting the rights of all animals.

Where to Find Organic and Sustainable Products

Cowgirl Creamery—www.cowgirlcreamery.com
Truly Old World artisans in the craft of cheesemaking, Peggy Smith and Sue Conley make some of the most respected and award-winning organic cheeses in the U.S. from their renovated barn in Point Reyes Station, California.

Generation Green—www.generationgreen.org
Generation Green gives families a voice in public policy decisions. As consumers, we have the power to reject corporate policies that endanger us and our children. Generation Green also produced the book *Fresh Choices: More Than 100 Easy Recipes for Pure Food When You Can't Buy 100% Organic*, to give people the ability to cook healthier meals without giving up flavor, or being inconvenienced by too many exotic and hard-to-find ingredients.

Heritage Foods USA—www.heritagefoodsusa.com
Heritage Foods USA exists to promote genetic diversity, small family farms, and a fully traceable food supply.

Native Seeds/SEARCH—www.nativeseeds.org
NS/S works to preserve knowledge about the traditional uses of the crops we steward. Through research, seed distribution, and community outreach, NS/S seeks to protect biodiversity and to celebrate cultural diversity. You can also order seeds for growing traditional Tohono O'odham tribal foods.

Niman Ranch—www.nimanranch.com
Niman Ranch started business nearly thirty years ago in Marin County, just across the Golden Gate Bridge from San Francisco. They work with over 300 independent family farmers across the U.S. who raise livestock for the company according to their strict protocols.

Seeds of Change—www.seedsofchange.com
Seeds of Change started in 1989 with a simple mission: to help preserve biodiversity and promote sustainable, organic agriculture by cultivating and disseminating an extensive range of open-pollinated, organically grown, heirloom and traditional vegetable, flower, and herb seeds.

Skagit River Ranch—www.skagitriverranch.com
Skagit River Ranch is a small family-owned certified organic farm in Washington state. George and Eiko Vojkovich (featured in this book) dedicate themselves to providing customers with the most healthy organic beef, chicken, and eggs available directly from the farm.

BOOKS, PERIODICALS, AND DVDS

Animal Liberation, by Peter Singer (Ecco, 2001)
First published in 1975, this book by bioethicist Peter Singer exposes the chilling realities of today's "factory farms" and product-testing procedures. It offers sound, humane solutions to what has become a profound environmental and social as well as moral issue.

Disease-Proof Your Child: Feeding Kids Right, by Joel Fuhrman, M.D. (St. Martins Press, 2005)
Dr. Fuhrman explains how eating particular foods (and avoiding others) can have a significant impact on your child's resistance to dangerous infections and their intelligence and success in school.

Eat Here: Homegrown Pleasures in a Global Supermarket, by Brian Halweil (W. W. Norton, 2004)
Brian Halweil, a senior researcher at the Worldwatch Institute, writes on the social and ecological impacts of how we grow food.

Fast Food Nation: The Dark Side of the All-American Meal, by Eric Schlosser (Perennial Books, 2002)
Fast Food Nation is a groundbreaking work of investigation and cultural history that is changing the way America thinks about the way it eats.

Food Politics: How the Food Industry Influences Nutrition and Health, by Marion Nestle (University of California Press, 2003)
Dr. Marion Nestle vividly illustrates food politics in action: watered-down government dietary advice, schools pushing soft drinks, diet supplements promoted as if they were First Amendment rights. When it comes to the mass production and consumption of food, strategic decisions are driven by economics—not science, not common sense, and certainly not health.

The Food Revolution: How Your Diet Can Help Save Your Life and Our World, by John Robbins (Conari Press, 2001)
John Robbins, the man who started the "food revolution" with the million-selling *Diet for a New America,* boldly posits that, collectively, our personal diet can save ourselves and the world.

The Future of Food—www.thefutureoffood.com
The Future of Food, a feature-length documentary produced by Deborah Koons Garcia, offers an in-depth investigation into the disturbing truth behind the unlabeled, patented, genetically engineered foods that have quietly filled U.S. grocery store shelves for the past decade.

Hope's Edge: The Next Diet for a Small Planet, by Frances Moore Lappé and Anna Lappé (Jeremy P. Tarcher, 2003)

Thirty years ago, Frances Moore Lappé started a revolution in the way Americans think about food and hunger. With this book, Frances and her daughter, Anna, pick up where *Diet for a Small Planet* left off.

New Vegetarian Baby, by Sharon Yntema (McBooks Press, 1995)
Incorporating all the latest information, this book will bolster your own instincts, answer your questions, and lay safely to rest any lingering doubts about a vegetarian regimen for infants.

The Vegetarian Sourcebook: Basic Consumer Health Information about Vegetarian Diets, Lifestyle, and Philosophy (Omnigraphics, 2002)
This excellent reference, packed with compelling statistics, describes various types of vegetarian diets and gives practical advice for safely incorporating them into everyday life.

VegNews—www.vegnews.com
This popular magazine focuses on vegetarian concerns, offering its over 100,000 readers up-to-date information on living a compassionate and healthy lifestyle.

SPECIAL CAUSES

Campaign to Label Genetically Engineered Foods—
www.thecampaign.org
Concerned with the growing acreage of unlabeled and inadequately tested genetically engineered crops, the Campaign to Label Genetically Engineered Foods was launched in March 1999.

Percy Schmeiser, Farmer/Activist—www.percyschmeiser.com
If you can assist financially with the legal challenge facing Percy, please visit his Web site or send your contribution to Fight Genetically Altered Food Fund Inc., Box 3743, Humboldt, SK, S0K 2A0, Canada.